STRENGTHENING THE LINKAGES BETWEEN THE SCIENCES AND THE MATHEMATICAL SCIENCES

Committee on Strengthening the Linkages Between the Sciences and the Mathematical Sciences

Commission on Physical Sciences, Mathematics, and Applications

National Research Council

NATIONAL ACADEMY PRESS
Washington, D.C.

NOTICE: The project that is the subject of this report was approved by the Governing Board of the National Research Council, whose members are drawn from the councils of the National Academy of Sciences, the National Academy of Engineering, and the Institute of Medicine. The members of the committee responsible for the report were chosen for their special competences and with regard for appropriate balance.

The material in this report is based on work supported by the National Science Foundation and the Defense Advanced Research Projects Agency under Grant No. DMS-9703610 and by DHHS Contract No. N01-OD-4-2139 between the National Academy of Sciences and the National Institutes of Health, an agency of the U.S. Department of Health and Human Services. Any opinions, findings, conclusions, or recommendations expressed in this publication are those of the author(s) and do not necessarily reflect the views of the organizations or agencies that provided support for the project.

International Standard Book Number 0-309-06947-5

Additional copies of this report are available from:

Commission on Physical Sciences, Mathematics, and Applications
National Research Council, NAS 285
2101 Constitution Avenue, N.W.
Washington, DC 20418
202-334-3061

Copyright 2000 by the National Academy of Sciences. All rights reserved.

Printed in the United States of America

THE NATIONAL ACADEMIES

National Academy of Sciences
National Academy of Engineering
Institute of Medicine
National Research Council

The **National Academy of Sciences** is a private, nonprofit, self-perpetuating society of distinguished scholars engaged in scientific and engineering research, dedicated to the furtherance of science and technology and to their use for the general welfare. Upon the authority of the charter granted to it by the Congress in 1863, the Academy has a mandate that requires it to advise the federal government on scientific and technical matters. Dr. Bruce M. Alberts is president of the National Academy of Sciences.

The **National Academy of Engineering** was established in 1964, under the charter of the National Academy of Sciences, as a parallel organization of outstanding engineers. It is autonomous in its administration and in the selection of its members, sharing with the National Academy of Sciences the responsibility for advising the federal government. The National Academy of Engineering also sponsors engineering programs aimed at meeting national needs, encourages education and research, and recognizes the superior achievements of engineers. Dr. William A. Wulf is president of the National Academy of Engineering.

The **Institute of Medicine** was established in 1970 by the National Academy of Sciences to secure the services of eminent members of appropriate professions in the examination of policy matters pertaining to the health of the public. The Institute acts under the responsibility given to the National Academy of Sciences by its congressional charter to be an adviser to the federal government and, upon its own initiative, to identify issues of medical care, research, and education. Dr. Kenneth I. Shine is president of the Institute of Medicine.

The **National Research Council** was organized by the National Academy of Sciences in 1916 to associate the broad community of science and technology with the Academy's purposes of furthering knowledge and advising the federal government. Functioning in accordance with general policies determined by the Academy, the Council has become the principal operating agency of both the National Academy of Sciences and the National Academy of Engineering in providing services to the government, the public, and the scientific and engineering communities. The Council is administered jointly by both Academies and the Institute of Medicine. Dr. Bruce M. Alberts and Dr. William A. Wulf are chairman and vice chairman, respectively, of the National Research Council.

COMMITTEE ON STRENGTHENING THE LINKAGES BETWEEN THE SCIENCES AND THE MATHEMATICAL SCIENCES

THOMAS F. BUDINGER, E.O. Lawrence Berkeley National Laboratory, *Chair*
R. STEPHEN BERRY, University of Chicago
NICHOLAS R. COZZARELLI, University of California, Berkeley
INGRID DAUBECHIES, Princeton University
MARTIN FARACH-COLTON, Rutgers, The State University of New Jersey
LARS PETER HANSEN, University of Chicago
FRANK CHARLES HOPPENSTEADT, Arizona State University
JOHN P. LEHOCZKY, Carnegie Mellon University
GREGORY J. McRAE, Massachusetts Institute of Technology
MERCEDES PASCUAL, Center of Marine Biotechnology, University of Maryland
  Biotechnology Institute
JOSEPH PEDLOSKY, Woods Hole Oceanographic Institution
JAMES L. PHILLIPS, The Boeing Company
I.M. SINGER, Massachusetts Institute of Technology
MICHAEL TABOR, University of Arizona
SUZANNE DAVIES WITHERS, University of Washington
ELLEN GOULD ZWEIBEL, University of Colorado, Boulder

STACEY BURKHARDT, Project Assistant (through March 1998)
HEIDI L. DAVIS, Senior Program Officer (through January 1999)
NORMAN METZGER, Executive Director, CPSMA (through November 1999)
LA VONE WELLMAN, Project Assistant (through March 1999)
DOROTHY ZOLANDZ, Study Director

COMMISSION ON PHYSICAL SCIENCES, MATHEMATICS, AND APPLICATIONS

PETER M. BANKS, Veridian ERIM International, Inc., *Co-chair*
W. CARL LINEBERGER, University of Colorado, *Co-chair*
WILLIAM F. BALLHAUS, JR., Lockheed Martin Corporation
SHIRLEY CHIANG, University of California, Davis
MARSHALL H. COHEN, California Institute of Technology
RONALD G. DOUGLAS, Texas A&M University
SAMUEL H. FULLER, Analog Devices, Inc.
JERRY P. GOLLUB, Haverford College
MICHAEL F. GOODCHILD, University of California, Santa Barbara
MARTHA P. HAYNES, Cornell University
WESLEY T. HUNTRESS, JR., Carnegie Institution
CAROL M. JANTZEN, Westinghouse Savannah River Company
PAUL G. KAMINSKI, Technovation, Inc.
KENNETH H. KELLER, University of Minnesota
JOHN R. KREICK, Sanders, a Lockheed Martin Company (retired)
MARSHA I. LESTER, University of Pennsylvania
DUSA M. McDUFF, State University of New York at Stony Brook
JANET L. NORWOOD, Former Commissioner, U.S. Bureau of Labor Statistics
M. ELISABETH PATÉ-CORNELL, Stanford University
NICHOLAS P. SAMIOS, Brookhaven National Laboratory
ROBERT SPINRAD, Xerox PARC (retired)

NORMAN METZGER, Executive Director (through July 1999)
MYRON F. UMAN, Acting Executive Director

Preface

The relationship between the sciences and the mathematical sciences is a long and symbiotic one. The two have grown up together, repeatedly interacting, with discoveries in science opening up new problems in mathematical science and advances in mathematics, statistics, computer science, and operations research enabling new practical technologies and advancing entirely new frontiers of science.

The cooperation and collaboration of mathematical scientists with engineers and scientists have occurred through chance encounters, and more often than not a scientific or technological problem exists for years before a mathematician or statistician recognizes or discovers the problem as interesting and mathematically tractable. Although a biologist, medical scientist, geophysicist, or economist might naturally seek out a mathematical scientist to embark on a joint research venture, collaborations are frequently impeded by hurdles ranging from a dissimilarity in scientific language and understanding to a lack of funding. On the one hand, cultural and institutional barriers can stand in the way of training mathematical scientists in interdisciplinary research, and on the other, educational priorities can limit the exposure of students in other sciences to mathematics curricula. Faculty interested in and encouraged to pursue cross-disciplinary research can face barriers associated with precedents and the attitudes of some of their colleagues. The success of many individual partnerships and that of the few organized institutes that encourage and fund interdisciplinary ventures give ample evidence that cross-disciplinary research and education can bring about great social benefits. In addition, several recently published analyses have recommended that the United States should strengthen its mathematical science resources or risk losing its preeminent position in international research.

In response to these concerns, and at the request of several federal agencies, the National Research Council established the Committee on Strengthening the Linkages Between the Sciences and the Mathematical Sciences. The charge to the committee was

> . . . to examine the mechanisms for strengthening interdisciplinary research between the sciences and mathematical sciences, with the principal efforts of the committee being to suggest what are likely to be the most effective mechanisms for collaboration, and to implement them through the Internet, widely circulated reports, and other dissemination activities, such as campus workshops convened by committee members. The committee will also examine implications for education in the sciences and mathematics and suggest changes in graduate training intended to reinforce efforts to strengthen the dialogue among the sciences.

The committee was composed of researchers and educators from various disciplines who were themselves known for their interest in and pursuit of cross-disciplinary research at the math/science interface. The committee met five times over the course of a year, both to receive testimony from others engaged in conducting or supporting cross-disciplinary research and to discuss its members' own experiences and knowledge of cross-disciplinary pursuits. Previous studies and recommendations related to cross-disciplinary research and enhancement of the mathematical sciences were reviewed. The current cross-disciplinary activities of government bodies and university scientists were examined as models for successful collaboration.

It is the committee's sincere hope that this report will motivate both young and established researchers and educators to pursue training and research in cross-disciplinary areas and that it will encourage academic institutions to create mechanisms that foster cross-disciplinary education and research and that reward faculty appropriately. The committee hopes the recommendations will serve as guidance for funding agencies, both government and private, to advance research in the cross-disciplinary areas relevant to their missions and that it will demonstrate to the general reader the importance of such interactions.

Thomas F. Budinger, *Chair*
Committee on Strengthening the Linkages Between
the Sciences and the Mathematical Sciences

Acknowledgment of Reviewers

This report has been reviewed by individuals chosen for their diverse perspectives and technical expertise, in accordance with procedures approved by the Report Review Committee of the National Research Council (NRC). The purpose of this independent review is to provide candid and critical comments that will assist the authors and the NRC in making the published report as sound as possible and to ensure that the report meets institutional standards for objectivity, evidence, and responsiveness to the study charge. The contents of the review comments and draft manuscript remain confidential to protect the integrity of the deliberative process. The committee wishes to thank the following individuals for their participation in the review of this report:

Elwyn R. Berlekamp, University of California, Berkeley,
Alexandre J. Chorin, University of California, Berkeley,
Kenneth A. Dill, University of California, San Francisco,
Ronald G. Douglas, Texas A&M University,
Robert W. Dutton, Stanford University,
L.B. Freund, Brown University,
Ernest M. Henley, University of Washington,
William G. Howard, Jr., independent consultant,
Harry Kesten, Cornell University,
Nancy J. Kopell, Boston University, and
Shmuel Winograd, IBM T.J. Watson Research Center.

Although the individuals listed above have provided many constructive comments and suggestions, responsibility for the final content of this report rests solely with the authoring committee and the NRC.

Contents

Executive Summary		1
1	Introduction	5
	References	7
2	Examples of Linkages	9
	Cross-Disciplinary Efforts by Researchers	9
	Lessons Learned	14
	Programs That Foster Cross-Disciplinary Research	17
	Success Factors	27
	References	28
3	Previous Efforts and Studies Relevant to Math/Science Linkages	31
	References	33
4	Recommendations	35
	Funding Directions	35
	Cross-Disciplinary Interactions	38
	Oversight	39
	Reference	41
Appendixes		43
A	Ten Case Studies of Math/Science Interactions	41
	1 Modeling Weather Systems Using Weakly Nonlinear, Unstable Baroclinic Waves	45
	2 Mixing in the Oceans and Chaos Theory	47
	3 Wavelets: A Case Study of Interaction Between Mathematics and the Other Sciences	50
	4 From Forest Dynamics to Interacting Particle Systems	54
	5 Modeling the Dynamics of Infectious Diseases: Two Examples	57
	6 Topology and Dynamics of Mutant Bacteria, and Applications to Materials Science	60

	7	Examples from Molecular Biology	63
	8	Challenges, Barriers, and Rewards in the Transition from Computer Scientist to Computational Biologist	67
	9	Crossbreeding Mathematics and Theoretical Chemistry: Two Tales with Different Endings	69
	10	Martingale Theory	71
B		Workshop Agenda and Presentations	75
C		Partial Chronology of Previous Efforts to Strengthen Mathematics and Cross-Disciplinary Research	91
D		Federal Agencies That Provide Funding Opportunities	101
E		Acronyms and Abbreviations	121

STRENGTHENING THE LINKAGES
BETWEEN THE
SCIENCES
AND THE
MATHEMATICAL SCIENCES

Executive Summary

Mathematics and the sciences[1] have grown up together, repeatedly interacting. Discoveries in science open up new advances in statistics, computer science, operations research, and pure and applied mathematics. These, in turn, enable new practical technologies and advance entirely new frontiers of science. Frequently, however, cooperation and collaboration between mathematical scientists and scientists have come as a result of chance encounters. Often a scientific or technological problem exists for years before a mathematical scientist discovers it and recognizes it as interesting and mathematically tractable. Similarly, scientists have frequently expended energy "inventing" mathematical solutions that existed for decades but were not familiar to them from their training.

To encourage new linkages between the mathematical sciences and other fields and to sustain old linkages, the National Research Council (NRC) constituted a committee representing a broad cross-section of scientists from academia, federal government laboratories, and industry. Its task was to "examine the mechanisms for strengthening interdisciplinary research between the sciences and mathematical sciences, with the principal efforts being to suggest what are likely to be the most effective mechanisms for collaboration . . . ."

The committee believes that the benefits to be derived from cross-disciplinary[2] activities linking the mathematical sciences with science and engineering are indeed huge and far reaching. Its recommendations can be found in Chapter 4. The main elements of these recommendations are summarized here.

---

[1] The terms "mathematics" and "science" are used in their broadest sense: for the purposes of this report, "mathematics" means the mathematical sciences, which include pure mathematics, applied mathematics, statistics and probability, operations research, and scientific computing; "science" includes the biomedical sciences, engineering sciences, computer science, life sciences, physical sciences, and social and behavioral sciences.

[2] For the purposes of this report, "cross-disciplinary" denotes interactions between the sciences and the mathematical sciences.

FUNDING DIRECTIONS

The committee recommends that additional funding be allocated to initiatives that will strengthen existing linkages between the mathematical sciences and other sciences and that will build new linkages.

New resources are required if the nation is to realize the enormous potential of cross-disciplinary fertilization. However, a key premise of the committee's report is that the basic sciences and the mathematical sciences must remain healthy if cross-disciplinary research is to advance. Some disciplines will, accordingly, need additional funds to be capable of developing cross-disciplinary ties. Funding for cross-disciplinary research must not compromise support for basic disciplinary research or for individual investigators.

Chapter 4 contains detailed recommendations to funding agencies that will enhance multidisciplinary activities: support for summer institutes, workshops at existing research centers, and new science and technology centers. The committee also recommends fellowship programs of various types to sustain research scientists in their pursuit of compelling multidisciplinary ideas.

Case studies of math-science research linkages provide compelling evidence for the synergism between science and mathematics. They also elucidate the factors that made these cross-disciplinary efforts possible, as well as the barriers that inhibited them. Appendix A presents the 10 case studies and Chapter 2 discusses the lessons learned from them, with an emphasis on the obstacles to cross-disciplinary collaborations. Some obstacles are resource-related, and it is these obstacles the committee hopes to overcome with its recommendation on funding. Other obstacles are cultural: different disciplines have different goals, philosophies, and languages.

The committee found that many institutions have overcome barriers to cross-disciplinary research despite the obstacles and have flourished. Examples of such programs can also be found in Chapter 2. Moreover we can anticipate even more activity at the interface of mathematics and other fields in the future. Universities with vision are developing, at an increasing pace, new interdisciplinary programs that cut across traditional departmental and college boundaries; such programs are vital to the strategic planning of a university's educational and research mission. Mathematics can, because of its uniquely central role in science and engineering, play an important part in these plans for interdisciplinary growth.

CROSS-DISCIPLINARY INTERACTIONS

The committee recommends that academic institutions take responsibility for implementing vigorous cooperative programs between the sciences and the mathematical sciences.

As the examples show, departments and their faculty can actively pursue cross-disciplinary research opportunities and make changes in the curriculum that will give students cross-disciplinary skills. It is particularly important to develop effective criteria for the evaluation of cross-disciplinary research to ensure that promotion, tenure, and reward mechanisms fully recognize the importance of quality cross-disciplinary research.

OVERSIGHT

The committee recommends that a new standing committee be established with a long-term focus on improving the linkages between the mathematical sciences and other sciences in both academia and industry.

A standing committee would have as its special mission the fostering and nurturing of such linkages by monitoring and evaluating successful initiatives and by communicating its findings to the community. The committee could advise funding agencies on how they might better evaluate cross-disciplinary research, advise private foundations on effective methods for supporting such research, and inform professional societies about multidisciplinary research opportunities and related educational opportunities. Some specific responsibilities for such a committee are suggested in Chapter 4.

1

Introduction

Mathematics and the sciences have developed hand in hand as man has sought to understand the physical universe surrounding him. Over three hundred years ago, Galileo is reported to have said, "The laws of nature are written in the language of mathematics." Nobel laureate Eugene Wigner entitled his Courant lecture "The Unreasonable Effectiveness of Mathematics in the Natural Sciences" (Wigner, 1960). He emphasized that mathematical concepts turn up in entirely unexpected natural applications and that the enormous usefulness of mathematics borders on the mysterious.

Many mathematical applications are *not* mysterious. Mathematics has developed out of attempts to explain the world around us: for example, the calculus, differential equations, the Fourier series and the Fourier transform, and statistics all arose out of scientific problems. Not only has the application of these methods advanced engineering and the sciences, but the methods have also become an integral part of mathematics. Sometimes, moreover, mathematics gained insights on its own that greatly influenced other fields: the importance of symmetry groups in coding theory, chemistry, and physics and of non-Euclidean and differential geometry in relativity and string theory.

In this report the Committee on Strengthening the Linkages Between the Sciences and the Mathematical Sciences will highlight a few examples (among many) of the effective interaction between mathematics and other fields: the beginning of modern weather prediction, the development of biostatistics, mathematical economics, and biomedical imaging. Behind these dramatic developments lies a vast array of research activities at the interface of mathematics and other disciplines.

With the advent of the computer, linking mathematics and science is even more imperative. Mathematical modeling has become a third investigative approach in many areas of science, providing an important complement to the classical investigative approaches of theory and experimentation. In a number of sciences, modeling and simulation are the only viable complement to theoretical studies, because many problems cannot be addressed experimentally.

Large volumes of quantitative data with increasingly rich structures necessitate new mathematical approaches for their analysis. The evaluation of astronomical, geophysical, agricultural, climate, weather, economic, and genomic data now involves addressing data sets with one million times more data (i.e., tera- and petabyte data sets) than were previously involved in analysis. The day-to-day experiments in human gene expression now require techniques utilizing concepts not of 3 dimensions but of 600 dimensions.

Performing large-scale computations on complex systems and visualizing and analyzing the results pose new mathematical problems. The solutions to these problems will require new engineering science developments in computer architectures and communications and visualization systems. Future generations of computers will challenge mathematics even more.[1]

The need for research linkages is well described in *Science, Technology, and the Federal Government: National Goals for a New Era* (NAS, NAE, and IOM, 1993):

> Traditionally, science has been organized into specific disciplines. However, science, by its nature, is in continual flux. New disciplines emerge at the edges or intersections of existing ones. Old disciplines are transformed by new knowledge and new techniques, while new disciplines draw knowledge and techniques from the old.
>
> Furthermore, many of the problems that scientists are now trying to solve require contributions from more than one discipline. For this interdisciplinary research to succeed, scientists must be able to extend their knowledge to new areas and work effectively as members of teams.
>
> The performers and funders of research must allow these dynamics of science to drive its organization. They must remove barriers to emerging areas of research and encourage permeable institutional structures that allow for the flow of interdisciplinary opportunities.

To encourage new linkages between mathematics and other fields and to sustain old ones, the NRC appointed the committee and gave it the following charge:

> ... to examine mechanisms for strengthening interdisciplinary research between the sciences and mathematical sciences, with the principal efforts of the committee being to suggest what are likely to be the most effective mechanisms for collaboration, and to implement them through the Internet, widely circulated reports, and other dissemination activities, such as campus workshops convened by committee members [and to] examine implications for education in the sciences and mathematics and suggest changes in graduate training intended to reinforce efforts to strengthen the dialogue among the sciences.

The committee represents a cross-section of scientists and mathematicians from academia, national laboratories, and industry. The members' backgrounds encompass disciplines from the biomedical, life, physical, engineering, and social sciences, and from different mathematical sciences, including statistics and theoretical computer sciences. The methods used by the committee to arrive at the recommendations in this report include the following:

- Examination of case studies documenting the value of cross-disciplinary work involving mathematical sciences;

---

[1] For a few specific examples of research areas ripe for advancement through cross-disciplinary efforts, see the National Science Foundation (NSF) white paper "Mathematics and Science," by A. Chorin and M. Wright, available from the NSF Division of Mathematical Sciences.

- Personal experience and interviews with peers, administrators, representatives of industry, federal agencies, and private foundations;
- Examination of the mechanisms developed by different institutions to overcome some of the barriers to successful cross-disciplinary work;
- Consideration of related studies conducted by the NRC, federal agencies, and professional societies since 1980;
- Consideration of the status of a variety of efforts under way to enhance the opportunities and climate for cross-disciplinary work; and
- Debate among committee members leading to a consensus on the key issues and recommendations of this report.

The committee focused on the first part of its charge. Although it agreed early in the course of its work that it could not credibly propose meaningful reforms of graduate education, it nonetheless felt that such education is an extremely important topic that needs to be addressed. Top-down dictums meant to apply to all of science and/or mathematics and proposed by a group that by its nature had thin representation from some disciplines and none from others would at best be too general to be meaningful. Each community, discipline, or department responds to a specific set of both national and local needs and must determine for itself the role it should play in its community, in other words, its mission. Each organization's mission will largely determine the steps it can take at the graduate level to enhance cross-disciplinary linkages. Accordingly, this report relies instead on an anecdotal approach to provide ideas and inspiration. Those who recognize the value of cross-disciplinary linkages can use the lessons these anecdotes teach in formulating programs that meet their local needs.

Chapter 2 of this report discusses examples of cross-disciplinary research efforts; it distills those features that seem most likely to foster successful interactions and draws conclusions from them. The chapter also lists the major barriers to cross-disciplinary research. Chapter 3 looks at previous studies relevant to cross-disciplinary interactions and compares the findings and recommendations to the committee's own experience. Chapter 4 details the committee's recommendations. The committee believes the benefits to be accrued from activities linking the mathematical sciences with the other sciences could be huge and far-reaching. Successful linkages benefit both mathematics and science, and society as a whole will be greatly enriched.

Case studies of math-science interactions are presented in Appendix A, and a report on the workshop held by the committee to examine factors enabling and inhibiting cross-disciplinary research is presented in Appendix B. Appendix C gives a brief description of earlier studies that considered how to promote mathematics/science linkages. Appendix D is a beginner's guide to federal funding opportunities for projects linking the sciences and mathematical science. Appendix E lists acronyms and abbreviations.

REFERENCES

National Academy of Sciences (NAS), National Academy of Engineering (NAE), and Institute of Medicine (IOM). Committee on Science, Engineering and Public Policy. 1993. Science,

Technology, and the Federal Government: National Goals for a New Era. Washington, D.C.: National Academy Press.

Wigner, E. 1960. The unreasonable effectiveness of mathematics in the natural sciences. Communications on Pure and Applied Mathematics 13(February):1-14.

2

Examples of Linkages

This chapter summarizes the lessons the committee has learned from the 10 case studies of cross-disciplinary research compiled in Appendix A and the success factors that have been distilled from 11 examples of academic programs encouraging research and teaching at the math-science interface. The case studies of math-science research linkages provide compelling evidence of the synergism between science and mathematical sciences and the advances that can be made by such collaboration. They elucidate the factors that enable cross-disciplinary efforts as well as the barriers that inhibit them. The academic programs described in this chapter demonstrate the important role institutional structures play in the education of both undergraduate and graduate students and in the fostering of communication between mathematical scientists and other scientists.

CROSS-DISCIPLINARY EFFORTS BY RESEARCHERS

Boxes 2.1 to 2.4 give four brief examples of what can be achieved through successful math-science research linkages. Ten more-detailed case studies are set forth in Appendix A. Although they are not intended to be comprehensive, the four examples cut across the sciences and mathematical sciences and point out a variety of common features of math-science linkages.

All the case studies in this report share some common ingredients for success. In every case, mathematical scientists and scientists saw problems that were attractive and important within their own disciplines. Mathematical scientists and scientists had the opportunity to interact over long periods of time, generally in a common location. There was often an institutional structure, in some cases provided by a funding agency, to maintain the collaboration during its initial phase. The scientists and mathematical scientists shared elements of a common language, by virtue of broad educational backgrounds.

The case studies expose significant barriers to cross-disciplinary work. A young scientist's job search was hampered by the cross-disciplinary nature of her work. A renowned mathematician working on a scientific problem was told by his peers that he was wasting his time.

BOX 2.1 Biostatistics

The mathematical sciences have made significant contributions to many areas of science of special importance to mankind, and they, in turn, have been enriched by these contributions. One obvious example is the interactions between statistics and problems in medicine, epidemiology, and public health. These contributions have led to an entire subfield of the mathematical sciences, biostatistics, to address applications in these fields.

One of the most important individuals in the development of statistics was John Graunt. In 1662, Graunt turned his attention to the *Bills of Mortality*, weekly reports by London parish clerks giving the number and the causes of death. These had been instituted to help authorities detect the onsets of epidemics. Graunt published an analysis of these data and developed what is now known as a life table, which allows for the calculation of life expectancy. His work was especially ingenious in that he did not have basic population data from censuses to work with. The ideas and methods Graunt established more than 300 years ago have been highly refined and now form the basis for the life insurance industry and modern survival analysis.

The refinements of census methodology, especially in health and vital statistics, are one of the most important aspects of epidemiology, a body of methods designed to determine which group is likely to become ill, the reasons for illness, and what can be done to control an illness. An important method in medical science for determining the efficacy of treatments is the clinical trial. The initial formalization of the clinical trial method used the work of the famous statistician R.A. Fisher on randomization of treatments to comparable groups of experimental units. Fisher had developed these methods to obtain proper statistical tests of significance for agricultural variables such as soil type and fertilizer. With the inception of rigorous drug approval processes, the importance of clinical trials has intensified, as has the need for innovations to make the trials as informative and as efficient as possible. A variety of innovations have been introduced, for example sequential methods (see the case study on martingale theory in Appendix A). The methods developed to solve medical problems, such as determining the causes of certain diseases and evaluating the efficacy of various therapies, have become core elements of the discipline of statistics and have been applied to many other substantive areas.

Another important aspect of biostatistics involves the mathematical and statistical modeling of biological and biomedical phenomena. Formal models have contributed greatly to an understanding of the time course of epidemics, the carcinogenesis process at the cellular level, the dose-response behavior of animals or humans to drugs or toxic substances, the point processes of neuron firings, and the pharmacokinetics and pharmacodynamics of drugs in the bloodstream. Much of the early mathematical modeling involved relatively simple systems of differential equations or Markov process models. As the behavior of biological processes becomes known in much greater detail and more sophisticated scientific technologies are developed to measure biological processes, more complex mathematical models are required. For example, a new dimension of models is now being developed to gain an understanding of scientific phenomena such as protein folding, cognitive neuroscience, and genomics. These biological problems will ultimately create new branches of statistics and mathematics, and the mathematical and statistical sciences will ultimately help to improve our understanding of these areas.

SOURCE: Based on Colton and Armitage (1998); Johnson and Kotz (1982); and Stigler (1986).

BOX 2.2 The Charney-von Neumann Collaboration: Numerical Weather Prediction

Historically, weather forecasting relied on a set of intuitive experiential skills. Partly as a result of the impetus given to meteorology by the forecasting activities developed by the armed forces in World War II, forecasting was increasingly seen as a basic problem in fluid mechanics. Given the observed state of the atmosphere today, could one use the fundamental equations of fluid mechanics to calculate the state of the atmosphere tomorrow and the next day?

A major step forward on this problem was taken at the Institute for Advanced Study in Princeton shortly after the war, when the mathematician John von Neumann was looking for a scientific problem to which his emerging interest in computer computation could be successfully applied. He focused on meteorology and weather prediction as a problem whose needs could be made to fit with the foreseen capabilities of computer calculations at that time. It was Vladimir Zworykin, a scientist at RCA and friend of von Neumann's, who was originally interested in the meteorology problem. Von Neumann reached out to the meteorology community for support and organized a conference that was attended by the leading figure in meteorology at that time, Carl Gustav Rossby. Rossby suggested that the core of the group around von Neumann consist of young meteorologists interested in a mathematical approach to prediction, and the original funding proposal (to the Navy) called for a team with strong meteorological input.

Jule Charney, a young postdoctoral fellow at the time, was suggested to von Neumann and became the leader of the meteorology group. Von Neumann had a very deep knowledge of physics as well as mathematics. Charney, although a meteorologist, had trained as an undergraduate in math and had started graduate school in mathematics before switching fields. Charney and von Neumann were therefore free of many of the communication barriers that might otherwise exist between meteorologists and mathematicians. Charney believed that von Neumann's willingness to work outside the traditional mathematics domain was related to his European training, which had a strong tradition of mathematicians being very interested in physics. Nonetheless, von Neumann was criticized by his mathematical colleagues, who felt he was wasting his time in mundane concerns rather than contributing strictly to pure mathematics.

The key problem was how to approach the calculation of viscous fluid flow in response to pressure and pressure changes. The formal mathematics are contained in the Navier-Stokes equations, but these equations were clearly beyond the reach of the computers of the day. Even the so-called primitive equations in which the hydrostatic approximation is made are an enormous challenge since they contain fast-time-scale phenomena like gravity/acoustic waves largely irrelevant for the weather problem. Von Neumann's interest centered on the computational problem, and he was pushing a direct attack on the primitive equations. But Charney, the meteorologist, outlined a way of dealing with the original Navier-Stokes equations by a method now called singular perturbation theory to abstract a simplified, consistent, and powerful approximation to the Navier-Stokes equations that is today called the quasi-geostrophic approximation. And this was the path the Princeton group followed with success.

One can identify three principal ingredients for the success of the endeavor: (1) a sympathetic funding source, in this case the Navy Office of Research and Inventions, (2) a committed mathematician with a good grasp of physical principles and a leadership role in the mathematical side of the problem, and (3) scientists, in this case meteorologists, with a strong mathematical background prepared to profit from the interest of the mathematical community in their problem.

SOURCE: Based on Lindzen et al. (1980) and Platzman (1979).

BOX 2.3 Biomedical Imaging and Mathematics

The history of mathematics in biomedical imaging illustrates how mathematicians and speciality scientists can make rapid progress when they work in teams. Most of the early work on modern medical image reconstruction was developed very slowly by individuals working independently. At first, success was stifled by the lack of mathematical input, but later on, partnerships between mathematicians and medical scientists resulted in immediate successes.

The mathematical formulations underpinning the three-dimensional image reconstruction techniques now known as X-ray computer-assisted tomography (X-ray CT, also known as CAT scan), positron emission tomography (PET), single photon emission tomography (SPECT), and magnetic resonance imaging (MRI) were laid by Johann Radon in 1917, but the Radon transform was not discovered until 60 years later. The first success in reconstruction tomography involving elegant mathematical applications was that of physicist and radioastronomer Ronald Bracewell, who in 1956 used the Fourier projection (the central slice theorem) as the basis for reconstructing the regions of microwave radiation emitted from the Sun disk. The connection between Radon's mathematics and Bracewell's early work was not made until 20 years later, in the mid-1970s. The development of medical reconstruction tomography proceeded independently of Bracewell's contributions.

Medical computed tomography began to be developed in the early 1960s and proceeded slowly because there was little mathematical input. The earliest X-ray CT demonstration was by a neurologist, William Oldendorf, who in 1961 single-handedly engineered an X-ray reconstruction of the transverse section of an object consisting of iron and aluminum nails. Although an inventive experimental study, it utilized a crude method of simple back projection. The patented invention that resulted was deemed impractical because it required lengthy analysis. Oldendorf worked without the input of a mathematician and without any knowledge of the work of Radon or Bracewell. In 1963, David Kuhl, a physician, and Roy Edwards, an engineer, invented a method of imaging radionuclide distributions. They even performed clinical studies in patients 9 years before the first patient X-ray tomogram. Since the mathematics needed for an accurate mapping had not been incorporated into their method and computer operating systems in 1963 were unable to quickly perform even simple back projection, the resolution of Kuhl's scanner was only as good as that obtained with existing methods of radionuclide imaging.

A crucial mathematical contribution to the reconstruction problem was made in 1963 and 1964 by the physicist/mathematician Allan Cormack. His contributions were directly motivated by two problems. First, medical radiotherapy treatment required the ability to determine the body's attenuation coefficient distributions so that externally applied radiation could be targeted at the tumor. Second, a mathematical algorithm was needed for reconstructing the three-dimensional distribution of radionuclide concentrations from data collected by a PET instrument developed in 1962.

Independent of the above developments, Godfrey Hounsfield, a computer engineer and industrial researcher, invented the first practical device for performing X-ray computer-assisted tomography on human beings. In 1967, oblivious of the earlier mathematics of Radon, Bracewell, or Cormack and of the instruments developed by Oldendorf and Kuhl, Hounsfield used an X-ray source and X-ray detector with a test bed for obtaining projections through a cadaver brain. Nine hours of data acquisition and two hours of computation were required to obtain a single two-dimensional plane from multiple one-dimensional profiles or projections. Hounsfield used a simple arithmetic reconstruction technique that was merely an iterative estimation method of solving a series of simultaneous equations (i.e., each equation represented a line integral of attenuation coefficients through the head). This method was entirely independent of and different from the method published 8 years earlier by Cormack.

Eventually, both Cormack and Hounsfield were recognized with a Nobel Prize for their contributions to the development of medical imaging. But it is clear that the ideas and mathematics were independently discovered by these and other scientists mentioned in this example. The key to the application of computed tomography in hospitals was the computer. In 1970, when practical disc operating systems first became available, it was immediately recognized that by using the fast Fourier transform and algorithms based on the work of Radon, Bracewell, and Cormack, a practical medical X-ray CAT scanner could be manufactured.

In the early 1970s, a number of mathematician-scientist pairs commenced a series of discoveries that led to modern medical three-dimensional imaging. Noteworthy contributions circa 1974 came from three teams. First there were the contributions of Lawrence Shepp to the practical implementation of reconstruction. That work was clearly the result of his partnership with physicist Jerome Stein and physician Sadek Hilal in the filtered back projection method of reconstruction now used. Then, the team of Robert Marr, Paul Lauterbur, and Lawrence Shepp showed the power of arithmetic and Fourier techniques in three-dimensional reconstruction from projections in MRI. Grant Gullberg, Ronald Huesman, and Thomas Budinger, another team of mathematician, physicist, and physician, showed solutions to the attenuated Radon transform problem in SPECT. Currently, MRI and SPECT are being applied by teams of mathematicians, computational scientists, and statisticians to problems ranging from earthquake prediction to understanding and treating mental disorders, heart disease, and cancer.

The message from this historical synopsis is that mathematicians and scientists working in separate locations and on seemingly unrelated scientific objectives related to the mathematical inverse problem made slow and generally unrecognized progress. But when both mathematicians and scientists worked together, as did the three teams cited above, progress was rapid and almost immediately significant.

SOURCE: Based on Herman (1979); Natterer (1986); and Deans (1983).

BOX 2.4 Economics and Game Theory

Modern game theory has provided economists with mathematical tools for investigating resource allocation conflicts between groups of adversarial agents ("players"). These agents can be firms competing for market share, governments vying for advantages in trade, or firms and workers bargaining over labor contracts. Mathematician John von Neumann and economist Oskar Morgenstern were founders of modern game theory, and their collaboration offers an intriguing illustration of the linkages between mathematics and sciences. Mathematician John Nash extended their work in ways that made game theory applicable to a rich collection of conflicts of vital interest to the field of economics.

Morgenstern and von Neumann's book, *Theory of Games and Economic Behavior*, was published in 1944. At the time, Morgenstern was in the Economics Department at Princeton and von Neumann was at the Institute for Advanced Study. Morgenstern was skeptical of the then preeminent role of the Keynesian paradigm because of its naive treatment of incentives and individual decision making. This skepticism was clearly evident in his earlier writings. It was his collaboration with von Neumann, however, that allowed Morgenstern to translate his skepticism into an alternative approach to economic modeling. While the formal results in the book were due to von Neumann, Morgenstern's perspective was vital for attracting the attention of economists. Indeed some economists were quick to recognize the potential importance of game theory to their discipline, although it would take decades before this view was widely held. Von Neumann was a

> polymath with a brilliant grasp of many categories of knowledge as well as mathematics. His previous interest in and exposure to economics undoubtedly smoothed communication between him and Morgenstern. In fact, von Neumann had already made contributions to economic dynamics before his collaboration with Morgenstern. While the work of the two men provided the impetus for the use of game theory in economics, its formal results were limited in scope.
>
> Morgenstern and von Neumann's work set the stage for John Nash, who made seminal contributions to game theory while he was a graduate student in mathematics at Princeton University in the late 1940s and early 1950s. In his PhD dissertation in mathematics, Nash developed formally the concept of an *n*-person game and an equilibrium point of that game. He established the existence of equilibria in this setting, extending von Neumann's existence result for two-player, zero-sum games. Thus, Nash's formulation was much richer and opened the way to the use of game theory in understanding many problems that are now central to the field of economics. In related work, Nash formally posed the notion of bargaining between two players and provided a set of axioms that rationalized a solution to the "bargaining problem." His papers showed how game theory could be used to study economic conflicts that formerly had been viewed as beyond the reach of standard formal economic analysis. What is intriguing about Nash is that his contributions, which eventually were recognized with a Nobel Prize in economics, were made while he was a graduate student in mathematics at Princeton. Part of his inspiration came from the von Neumann-Morgenstern book, which had been published just a few years before Nash's arrival at Princeton and had attracted considerable interest among mathematicians at Princeton and elsewhere. More important input came from a seminar in game theory that mathematicians Tucker (who later became Nash's advisor), Kuhn, and Gale were running at Princeton when Nash was a graduate student. Although von Neumann was initially dismissive of some of Nash's ideas, both Tucker and Gale were supportive. Neither, of course, could anticipate the eventual influence of Nash's work on economics and other social sciences. While it was Nash's cleverness and creativity that allowed him to make a major advance over von Neumann's existence result zero-sum games, the broad view of mathematics at Princeton and the rather indirect influence of the economist Morgenstern helped to foster this work.
>
> In 1994 Nash shared the Nobel Prize in economics with John Harsanyi and Reinhard Selten. All three researchers did important work on the theory of noncooperative games, but it was Nash's work that laid the foundations for this line of research. His work also preceded Harsanyi's and Selten's work by almost two decades. The two-decade time lag between Nash's work on game theory and the important extensions by Harsanyi, Selten, and others leads to speculation about what might have been gained by restructuring research environments in economics and mathematics. Would the integration of game theory into economic analysis have proceeded at a faster pace had there been a more interactive research environment for economists and mathematicians? The likely answer is yes. The critical roles of von Neumann and Nash in the initial development of game theory provide powerful support for this conjecture.
>
> SOURCE: Based on Gul (1997); Leonard (1995); and Nasar (1998).

LESSONS LEARNED

From the 10 case studies in Appendix A, the committee compiled a list of two kinds of factors: those that enable math-science collaborations and other cross-disciplinary pursuits and those that militate against them (Box 2.5). Next, the committee categorized impediments to cross-disciplinary research and education and elaborated on them: career-related obstacles, obstacles related to the research culture, and resource-related obstacles.

> **BOX 2.5 Lessons Learned from the 10 Case Studies**
>
> *Enabling Factors*
>
> - Researchers had enough science and mathematics background in common (case studies 1, 2, 3, 4, 8, 9).
> - The problems addressed in the collaboration were attractive to researchers in both disciplines (case studies 2, 4, 5, 8 ,9, 10).
> - Researchers worked at the same facility for a sufficient period of time (case studies 2, 3, 4, 6, 7, 8, 9).
> - The research problem attracted enthusiastic, young researchers (case studies 2, 3, 4, 7, 9).
> - Institutional structures existed to foster repeated and prolonged interactions between the collaborators (case studies 1 and 8).
> - The researchers received early encouragement or funding from agencies, often through the vision of a single federal program officer (case studies 2, 5, and 7).
>
> *Impeding Factors*
>
> - Conservatism inherent in the discipline led to an underappreciation of cross-disciplinary research (case studies 2 and 3).
> - Differences in jargon made it difficult for collaborators to formulate research problems usefully (case studies 1 and 7).
> - Collaborators faced uncertain employment prospects because their research was interdisciplinary (case studies 2 and 8).
> - It was difficult to evaluate cross-disciplinary work for purposes of publication, promotion, and funding (case studies 6 and 8).
> - Contact across disciplines was maintained for too short a period to achieve real collaboration; the time required to establish cross-disciplinary work is daunting (case studies 7 and 9).

Career-Related Obstacles

A career path involving cross-disciplinary research and teaching has risks, ranging from diminished recognition of teaching efforts to delayed or denied promotions. The criteria for hiring and promotion often do not credit or reward those considering cross-disciplinary work as highly as those who keep working within the discipline. Although some researchers might welcome this challenge, it could be a significant impediment to others in the early stages of their academic careers.

Cross-disciplinary research is time-consuming. The time invested learning another discipline or establishing a viable research collaboration means that the time from start-up to first publishable result can be significantly longer than for single-discipline research. This is particularly difficult for junior faculty on the tenure clock. Developing and teaching cross-disciplinary courses is also time-consuming. Even where tenure is not a factor, few departments have a good mechanism for assigning teaching credit for such courses or for advising students from outside the researcher's home department.

Cross-disciplinary research is not generally published in prestigious, discipline-oriented journals. This is particularly problematic for junior researchers, who generally must demonstrate contributions to the discipline via publication in respected journals. The multidisciplinary journals *Science, Nature,* and *Proceedings of the National Academy of Sciences* welcome cross-disciplinary papers, but they do not offer a large enough forum for the variety of cross-disciplinary research reports expected in the future. Even for these journals, it is harder to find qualified reviewers for cross-disciplinary research than for discipline-specific research, as few reviewers span fields.

The potential for diminished recognition can also deter collaborations. The academic structure rewards those who evince independent thinking and creativity, so the need to be a sole or the lead author to advance one's career is a fact of life. It is, as well, difficult to have balance in collaborative relationships. Research problems are usually perceived as either primarily science or primarily math problems, leading one researcher to be perceived as a lesser partner in the collaboration.

Cultural Obstacles

The culture of a discipline is such that it encourages and reinforces relationships between researchers within the discipline. Departmental meetings and seminars, professional society gatherings, and conferences bring together colleagues in a single discipline. This is an effective way to maintain the health of a discipline, but it also means that the typical researcher does not have much opportunity to meet and network with colleagues from other disciplines.

A lack of shared knowledge and a common professional language also inhibits collaborations. The case studies demonstrate that researchers may need to persist over a number of years in their attempts to communicate a problem to their colleagues in another discipline before there is a common appreciation of the essence of the problem. This may happen because the researchers do not yet understand one another's discipline well enough to fully grasp the depth of the problem, or it may happen simply because the potential collaborators do not understand each other's jargon. Potential collaborators need not be expert in each other's fields, but they must at least understand enough of the other discipline to recognize the contribution it can make to their research problem.

Resource Obstacles

At the level of the individual investigator, cross-disciplinary efforts by their nature generally require more time to start up and bear fruit than single-discipline efforts. This additional time generally translates into a need for more money to initiate and succeed at a cross-disciplinary research project than would be needed for a similar disciplinary project. This need for greater resources can be a barrier.

At the level of individual disciplines, a scarcity of human resources can limit the possibilities for cross-disciplinary research. For example, graduate student enrollment in mathematics departments has not kept pace with enrollment in the sciences or in other mathematical sciences. This dearth of human resources makes some mathematics departments, focused as they are on maintaining a critical mass of researchers to keep their basic discipline healthy, reticent to share their students by allowing them to participate in cross-disciplinary endeavors.

Across disciplines, the committee recognizes that some efforts are under way to move funding from the traditional disciplines to cross-disciplinary efforts. However, such efforts must be undertaken with caution. Some disciplines can benefit from shifting some of their resources from strictly disciplinary to cross-disciplinary activities. Others, such as mathematics, barely have the critical mass to support their core discipline and cannot benefit from such shifts. Some disciplines will, moreover, need additional monies to develop effective interdisciplinary ties.

PROGRAMS THAT FOSTER CROSS-DISCIPLINARY RESEARCH

The committee then examined some programs and institutes designed to foster linkages between the sciences and mathematical sciences. Here, too, it took a case study approach: the programs can be usefully examined to identify elements that contribute to their success and that can be emulated. The 11 examples cited below are not intended to be in any way exhaustive and were chosen largely because committee members had personal experience with their effectiveness.

The Mathematical Sciences Research Institute[1]

The Mathematical Sciences Research Institute (MSRI) of Berkeley, California, exists to further mathematical research through broadly based programs in the mathematical sciences and closely related activities. From its beginning, in 1982, MSRI has been primarily funded by the NSF, with additional support from other government agencies, private foundations, and academic and commercial sponsors. It is administered by a board of trustees drawn from academia, government laboratories, and the business world, with input from its committee of sponsoring institutions. These include about 35 of the leading mathematics research departments in the country, corporate sponsors, and several cooperating institutes and private foundations. A scientific advisory council of leading mathematical scientists oversees its scientific programs. MSRI's basic elements are mathematical programs and workshops and postdoctoral training.

Each year MSRI hosts between two and four major research programs covering a wide variety of topics in pure and applied mathematics. Many of these full-year and half-year programs involve researchers from the sciences, engineering, and commercial applied mathematics. Researchers interested in participating submit applications; they can apply to

---

[1]This description is drawn primarily from material on the MSRI Web site, <http://www.msri.org>.

participate in the full program or for 2 or 3 months. The MSRI program provides time, space, services, and at least partial salary to the researchers accepted. It also hosts 20 to 30 postdoctoral students each year for a semester or a year, bringing these students into contact with leaders in their fields of study. MSRI is successful at least in part because it brings together researchers from different disciplines for an extended period of time, in an environment that encourages interaction.

Each of the main programs also organizes an introductory workshop and one or more specialized workshops. These serve a broader community than the community of those who can come to MSRI for a longer period. In addition, MSRI hosts workshops on a variety of topics, from commercial applications of mathematics to the latest success in pure research. It also sponsors activities designed to develop human resources in the mathematical sciences, to communicate mathematics within and outside of the mathematics community, and to increase public awareness and appreciation of research in the mathematical sciences.

Large-scale programs at MSRI that directly involved other sciences include Mathematical Biology (1992), Strings in Mathematics and Physics (1991), Symplectic Geometry and Mechanics (1988-1989), and Mathematical Economics (1985-1986). More recently, there have been shorter events on financial mathematics, cryptography, physical oceanography, genomics, scientific imaging, and parallel computing. Programs planned for the near future include the mesoscopic structure of materials, computer-aided design, and quantum computing. MSRI recently established a subcommittee of its scientific advisory committee explicitly charged with identifying areas where collaborations between math and other sciences could be fruitfully encouraged and with developing program possibilities in these areas.

National Computational Science Alliance[2]

The National Computational Science Alliance (the Alliance) is a partnership between a number of institutions. Headquartered at the National Center for Supercomputing Applications (NCSA) at the University of Illinois at Urbana-Champaign (UIUC), it has as its primary goal fostering work in computational science and computational mathematics. The Alliance's efforts divide into four areas: high-performance computing resources, consulting, and training; enabling technologies development; applications technologies development; and educational applications.

Of particular interest are the Alliance's efforts in applications technology development. These efforts bring scientists, engineers, and computational scientists together to customize computational infrastructures for use by specific scientific and engineering communities. Efforts focus on six areas: chemical engineering, cosmology, environmental hydrology, molecular biology, nanomaterials, and scientific instruments. Work in each of these areas is carried out by teams whose members come from all parts of the United States. The teams are made up of scientists or engineers in the relevant scientific discipline and computational scientists and mathematicians. They are headed by a representative from the relevant scientific discipline and anchored by staff at NCSA who are assigned to the team. The goal in each area is to create computing infrastructure that can be used by the scientific community to attack important problems in various fields.

---

[2]Further information on the NCSA and the Alliance can be found at <http://www.ncsa.uiuc.edu>.

One example of the results of the Alliance's efforts is the Biology Workbench. This technology provides a point-and-click Web-based interface to more than 100 public domain databases of interest to molecular biologists and to software for sequencing DNA and identifying protein structures. The Biology Workbench frees biologists from needing to understand the various formats and syntaxes of individual databases in order to access the necessary information. It should give molecular biologists the ability to find, sort, and use the growing quantities of data generated by the community more efficiently and effectively than had been previously possible, allowing them to focus on their biology.

The interdisciplinary approach is a key theme of NCSA's adaptation of computational methods and is built into NCSA policy. Staff are alert to new possibilities for research linkages, and NCSA encourages facility users, Alliance partners, and staff to meet and talk across disciplinary lines and to consider novel research approaches. The interdisciplinary environment facilitates the transfer of technology and approaches between fields—for example, the molecular biology team adapted software developed by the nanomaterials team and applied it to the analysis of ion channels in cells.

The Institute for Mathematics and Its Applications, University of Minnesota[3]

The Institute for Mathematics and Its Applications (IMA) was established in 1982 with the mission of closing the gap between theories and their applications. This entails two tasks: (1) identifying problems and areas of mathematical research needed in other sciences and (2) encouraging the participation of mathematicians in these areas of application by providing settings conducive to the solution of such problems and by demonstrating that first-rate mathematics can make a real impact in the sciences.

The IMA scientific programs allow mathematicians and other scientists to share a stimulating research environment. Researchers who have spent time at IMA cite its success in building contacts between researchers of varied backgrounds and experiences; they also remark on the sense of community engendered there. Yearly programs are chosen for the purpose of encouraging interaction between mathematicians and scientists from academia, industry, and government laboratories and opening up new opportunities for the mathematical sciences. The topic of the year is usually divided into two or three subtopics, each of which the program concentrates on for 1 to 3 months and for which it holds a number of workshops. A typical yearly program is designed around a group of senior scientists who agree to be in residence for 3 to 10 months. This allows program continuity as well as scientific guidance for postdoctoral members. The IMA also runs a series of shorter programs during the summer. The annual programs have topics such as mathematics in biology, reactive flow and transport phenomena, mathematics in multimedia applications, and mathematics in the geosciences. Summer programs cover narrower topics, such as energy and environmental models for decision making under uncertainty, and codes, systems, and graphical models.

Postdoctoral fellows are a key component of IMA programs. Selected by open competition, they bring flexibility and enthusiasm to the program and challenge the senior members. Their participation is critical to the mission of the IMA, since it is expected that they

---

[3]This description is drawn primarily from material on the IMA Web site, <http://www.ima.umn.edu>.

will use their experience to become leaders in the mathematical-scientific community. Some of them work half of the time with industry and are supported by it.

Other IMA programs include the Industrial Problems Seminar, in which industrial scientists are invited to present industrial problems to IMA researchers; the IMA Participating Corporations Program, a formalized relationship between the IMA and industrial scientists; and IMA Participating Institutions, a consortium of universities that provides valuable support and guidance.

Research accomplishments have included mathematical advances in nonlinear waves, dynamical systems, and probability. The Industrial Problems Seminar alone generates about 10 published mathematical articles yearly.

Factors that contribute to IMA's success include (1) its ability to bring researchers together to interact for a period of weeks or months, (2) the presence of young researchers who have a less conservative view of research, and (3) the mentoring of postdoctoral fellows by senior researchers.

The Courant Institute of Mathematical Sciences[4]

New York University's Courant Institute of Mathematical Sciences is a leading center for research and graduate education in mathematics and computer science. Over the past 50 years, it has contributed to U.S. science by promoting an integrated view of the mathematical and computational sciences as a single, unified field. The Courant Institute has played a central role in the development of analysis, differential equations, applied mathematics, and computer science. Its research activity ranges from the theoretical to the applied. It covers a broad frontier that includes pure mathematics and computer science, as well as applications of mathematics and computation to the biological, physical, and economic sciences.

Richard Courant came to New York University in 1934 as a visiting professor, having left his position as director of the Mathematics Institute at the University of Göttingen, in Germany. In 1935, he was invited to build up the Department of Mathematics at the Graduate School of Arts and Science. In 1937, he was joined by Kurt O. Friedrichs and James J. Stoker. Together with a few of the faculty members already in the department, they formed a closely knit research group. During World War II, under the sponsorship of the Office of Scientific Research and Development, the team became the nucleus of an expanded group that undertook mathematically challenging problems arising from various war projects. However, in contrast to most other ad hoc teams, it did not abandon basic research and advanced instruction. After the war, support from the Office of Naval Research and other government agencies maintained the group and encouraged its growth. The name Institute for Mathematics and Mechanics was adopted in 1946. The Atomic Energy Commission installed a state-of-the-art electronic computer at New York University in 1952. This led to the creation of the Courant Mathematics and Computing Laboratory, which has functioned for many years under the auspices of the U.S. Department of Energy.

---

[4]This program description is derived primarily from the Courant Institute brochure, at <http://www.cims.nyu.edu/information/brochure>.

Central to the scientific life of the Courant Institute is its lively program of research seminars. Their purpose is to stimulate education and research at the level where the two are synonymous. Seminars promote the formation of working groups by drawing students and visitors into contact with ongoing research activities. They also keep the Courant community abreast of new developments around the world. In recent years, there have been regular seminars in applied mathematics, analysis, computational geometry, computer science, magnetofluid dynamics, probability and statistical physics, numerical analysis, programming languages, and robotics. Additional seminars are organized each year depending on the interests of the faculty and postdoctoral visitors; recent examples include mathematical biology, materials science, nonlinear waves, and turbulence.

Each year the Courant Institute is host to a large number of visiting scientists. Some are senior distinguished scientists on leave from their home institutions. Others are postdoctoral visitors with external support, typically from the NSF or a comparable foreign agency such as Canada's Natural Sciences and Engineering Research Council (NSERC), Italy's Consiglio Nazionale delle Ricerche (CNR), or France's Centre National de la Recherche Scientifique (CNRS). Still others are appointed to assist with one of the institute's many research projects. In addition, there are the Courant Institute instructorships and the Visiting Membership Program. Both serve to bring recent PhDs to the institute. These programs provide postdoctoral training and research support by involving young scientists in the institute's varied research activities. The programs have existed in one form or another since 1956, financed by various government agencies, industrial organizations, and private foundations. The NSF is currently the principal source of funding. Courant Institute instructorships are ordinarily for a 2-year term; they carry a teaching load of one course per semester. Visiting memberships are ordinarily for a 1-year term, but they carry no teaching duties; extension or renewal may be possible.

Current research efforts at Courant span a broad range of mathematics and applications. For example, a multidisciplinary effort has recently been launched to study the interactions of ice shelves with the atmosphere and ocean—an important unresolved scientific issue in current climate models. Simulation of complex biofluid dynamics phenomena, including blood flow through the heart, wave propagation along the basilar membrane of the inner ear, and the flight of insects, is being advanced using the immersed boundary method. The institute's recent focus on physiological neuroscience is complemented by research in computer vision. Work in computational physics includes projects in materials science, computational biochemistry, quantum mechanics, and electromagnetic scattering. This dynamic, cross-disciplinary institute is ranked among the top 10 mathematics departments nationwide (NRC, 1994).

The Courant Institute also follows a model of bringing a mix of senior and junior researchers together for an extended period of time, in a setting that provides many opportunities for interaction and mentoring. In addition to the long-term researchers, the steady flow of seminar speakers who come to the institute further stimulates the mix of ideas.

The University of Michigan Department of Mathematics[5]

The University of Michigan has a strong tradition of core mathematics research and training. However, in the early 1990s, department faculty determined that there was insufficient research and educational interaction with other departments to ensure the long-term health of the department. The department believed that both mathematics and the physical, biological, social, and management sciences flourish if there is genuine communication between them. Furthermore, it felt that improved educational cooperation would result from individual faculty contacts with colleagues from other departments. Department leadership believed that the best way to create such connections was by recruiting faculty whose research interests match those of colleagues from other science and engineering departments. At the urging of the department chair and with the support of the university administration, the department began its Interdisciplinary Initiative in 1993. Eight tenured faculty positions were committed to new faculty whose research interests cut across disciplinary boundaries. Four of these faculty positions have been filled.

The initiative has had a significant impact on the department and its cross-disciplinary interactions. A new and well-attended seminar in applied and cross-disciplinary mathematics with participants from a broad range of departments has been created. There are more joint research proposals being submitted and to a broader range of funding agencies. In addition, it has had a large impact on the curriculum. Several new courses at the graduate and advanced undergraduate level have been created, in addition to a new applied calculus honors sequence at the freshman-sophomore level and an undergraduate specialization in mathematical biology. Further, mathematics faculty recruited for the interdisciplinary initiative are playing a key role in the collaborative efforts of the department and the College of Engineering to review and possibly restructure the sophomore calculus sequence, the one taken by most engineering students. Also, a new doctoral program that encourages cross-disciplinary work is being set up.

Department of Statistics, Carnegie Mellon University[6]

The Carnegie Mellon University PhD program in statistics has a strong emphasis on cross-disciplinary training. The program is unusual in that training in cross-disciplinary applications of statistics is required of all students, along with training in probability and statistical theory and statistical computing.

Approximately half of the first-year curriculum is theoretical and half focuses on the practice of statistics. Some of the applied subject matter courses involve problems that arise from the faculty's collaborative research involvement. In the required course on statistical

---

[5] Information on the University of Michigan Department of Mathematics can be found at <http://www.math.lsa.umich.edu>. A detailed discussion of the success of this department can be found in AMS (1999).
[6] This program description is drawn from the paper "Modernizing Statistics Ph.D. Programs," written for the August 1993 symposium entitled Modern Interdisciplinary University Statistics Programs that was sponsored by the NRC's Committee on Applied and Theoretical Statistics. The full paper was reprinted in *American Statistician* 49 (February 1995):12-17.

practice, students gain experience in working on real applications. The applications cover a wide range of disciplines, and even though a student addresses only one or two applications, he or she learns about other applications from other students' experiences.

In the second year of the program, the advanced data analysis course exposes students to a substantial cross-disciplinary experience. The course has two main components that run throughout the year. One is a discussion of different types of data analysis problems, along with current tools and techniques. The second requires each student to engage in a major scientific research project and to analyze the data generated by that project. The student is supervised by a committee, one member of which is the scientist whose data is being analyzed. This culminates in a major report (which often becomes a research publication) and a presentation on the project to the entire department.

The third and fourth years of the PhD program focus on dissertation research; however, students also select a seminar course. A variety of such courses are offered. Some of them focus on landmark papers in statistics, some explore the latest developments in a subfield (such as Bayesian statistics, spatial statistics, or dynamic graphics), while others deal with application areas of faculty interest, such as statistics applied to cognitive neuroscience and neuro-imaging. Because among them the faculty have many cross-disciplinary research interests, students can also continue to gain a variety of additional cross-disciplinary research experiences by working as research assistants.

The primary intent of the graduate program is to train mathematical scientists for a variety of careers. PhD graduates conduct research in industrial and government laboratories, pursue research and teaching careers in traditional university settings, or pursue careers in the financial sector. The success of the program is due in large part to a strong faculty consensus on the importance of cross-disciplinary training. With knowledge increasing so rapidly, students may be frustrated by how little they can learn of it as graduate students, particularly if they are pursuing cross-disciplinary research. It is important, therefore, for them to receive consistent messages and good examples from the faculty concerning the importance of true cross-disciplinary experiences to their future success as statisticians.

University of Chicago Math Concentration of Specialization in Economics[7]

Undergraduate training of candidates for top PhD programs in economics by U.S. universities falls short of the training in European systems. Many foreign undergraduate students have substantially more mathematics training than U.S. students: they begin a specialized technical curriculum earlier and so have more opportunity to take courses beyond calculus, including real, complex, and functional analysis and ordinary differential equations.

In response to this situation, the University of Chicago has created an option that encourages college students who aspire to either a PhD in economics or a quantitative research position in private industry to take more mathematics. A math concentration with a specialization in economics incorporates seven quarter-long courses outside the Mathematics Department (intermediate economics courses, mathematical probability, statistics, and econometrics) with 10 courses from the undergraduate curriculum for a BS in mathematics.

---

[7]A description of the requirements of this specialization can be found at <http://www2.college.uchicago.edu/catalog99-00/htm/Math99.html>.

Courses from the Statistics Department are also included in the concentration. The program requires analysis and differential equations courses as well as two quarters of basic algebra.

This initiative came about because the directors of the undergraduate mathematics studies were willing to work with the undergraduate economics program directors, who recognized that quantitatively oriented students are likely to receive better training in mathematics from mathematicians than from economists. It brings technically strong mathematics students into a substantive science, in this case economics. Students with this mix of courses have gone on to graduate studies in economics and statistics. Others use this training to gain employment in research departments of private industries.

The University of Arizona Program in Applied Mathematics[8]

The Applied Mathematics Program at the University of Arizona was established 20 years ago by faculty from the Mathematics Department and other departments in the Colleges of Science and Engineering. Its mission is to provide graduate training leading to MS and PhD degrees and intellectual leadership in applied mathematics research and education at the university. The program recruits students from diverse undergraduate and master's backgrounds in the mathematical, physical, and engineering sciences. The students are supported through research assistantships, special fellowships, and teaching assistantships in the Mathematics Department.

The first year consists of basic mathematical coursework, and the student designs a program of study for subsequent years. Good faculty mentoring in an environment supportive of cross-disciplinary research is an important part of this process. Courses are offered by the Mathematics Department and other departments, and the program has also developed a number of uniquely cross-disciplinary training activities, such as an experimental teaching laboratory and a biomathematics initiative.

Students are ultimately able to apply a combination of mathematical sophistication and disciplinary knowledge to their chosen research area. They have produced dissertation research in such diverse areas as medical imaging, geoscience, atmospheric science, chemical engineering, aeronautical engineering, planetary science, neuroscience, molecular biology, and ecology and have gone on to research positions in academia, government laboratories, and industry.

There are approximately 50 active faculty members in the program from 15 departments in the colleges of Science, Engineering, and Medicine. The program falls under the jurisdiction of the Vice President for Research and the Graduate College, an administrative structure that helps it remain independent of any particular college. Although the Applied Mathematics Program is administratively and financially separate from the Mathematics Department, the two work together closely. This long-term, positive relationship has strengthened both units and is a key factor in the program's success.

---

[8] Further information on this program is available at <http://w3.arizona.edu/~applmath/home.shtml>. A detailed discussion of this program and its successes can be found in AMS (1999).

Spelman College Center for Scientific Applications of Mathematics[9]

The Center for Scientific Applications of Mathematics (CSAM) at Spelman College was launched 3 years ago to increase the number of African-Americans pursuing scientific careers by enhancing the scientific development of students at the undergraduate level. The primary objectives are to support research training and the development of stronger research programs at Spelman, as well as the professional development of faculty and students; to support curricular developments that incorporate interdisciplinary areas of study; and to develop partnerships with other educational, governmental, and industrial institutions that advance Spelman's scientific program.

CSAM attempts to break down the traditional barriers between academic departments. Its associated faculty sponsor a biweekly seminar to promote the sharing of research and curricular ideas across departmental lines. New seminars, courses, and course modules have emerged. The most significant results of crossing disciplinary barriers are the collegial connections that have been established and that lead to new collaborations.

Teams of students and faculty engage in research projects over a 12-month period, full-time during the summers. Approximately 60 percent of the full-time science and mathematics faculty have been involved in this aspect of the program. A visitor's program brings outstanding scientists to campus who can enlighten students on the interdisciplinary nature of today's scientific enterprise and share information on graduate programs and career paths. Through weekly seminars and presentations of their research at local, regional, and national conferences, students enhance their professional development.

Funding for this effort began when the W.K. Kellogg Foundation established CSAM as a Kellogg Center of Excellence in Science. Continuation funding was made available by Eastman Kodak. Factors that have been credited for the successful start of the program include (1) the vision and vitality of a senior mathematics faculty member who also has administrative responsibilities at the college, (2) a cooperative and dedicated faculty, and (3) a supportive senior administrative staff. CSAM provides a venue for faculty from different departments to interact repeatedly over time as they pursue the center's mission. Through CSAM, the college has begun publication of the *Spelman Science and Mathematics Journal,* an interdisciplinary undergraduate journal designed to enhance the technical writing skills of students and to increase communication on innovation in science education.

The Department of Energy Computational Science Graduate Fellowship[10]

The Department of Energy (DOE) supports a broad spectrum of basic and applied research in science and engineering at its national laboratories and through an extensive grants and contract program with universities and the private sector. High-performance computing is an integral part of DOE missions in climate change, biological systems, materials science, and national defense, as well as in many other areas. Accordingly, DOE has developed a program to encourage and sponsor graduate education in the application of mathematics and computational

---

[9] Based on a program description found in *Spelman Science and Mathematics Journal* 1(1). This regularly published journal can be found online at <http://www.spelman.edu/ssmj>.
[10] This program description is based on information found at <http://www.krellinst.org/CSGF>.

science to broad areas of scientific and technological development. The goal of the Computational Science Graduate Fellowship (CSGF) is to improve the quality and quantity of young scientists engaged in pursuits that depend on skills and understanding in mathematics and computing.

Predoctoral students at U.S. universities who are in their first or second year of graduate study in the physical, engineering, computer, mathematical, or life sciences may apply for the CSGF. The program provides tuition, an $1,800 per month stipend, an allowance for miscellaneous services, including travel, and computer purchase matching funds. Support can be for up to 4 years and must be renewed each year. The fellowship program requires a program of study that will provide a solid background in three areas: (1) a scientific or engineering discipline, (2) computer science, and (3) applied mathematics. A fellow's major field must fall in one of these categories, and the program of study must demonstrate breadth through substantial academic achievement in the other two. Submission of a program of study signifies a commitment to complete the courses listed if the fellowship is awarded. Changes in the program of study may be made only with the advance consent of the program's advisory committee.

A practicum (research assignment) at a DOE research laboratory is required of every fellow for at least one 3-month period during the term of the fellowship, with additional funding available to cover any extra expenses. The practicum is normally undertaken during the summer, as early as possible during the fellowship term.

Computational science graduate fellows become part of a group of researchers learning to solve problems outside disciplinary boundaries. A biennial conference brings participants together to share ideas and support one another. The most recent conference featured presentations on the use of computation in such areas as ecology, biomechanics, hydrology, signal processing, plasma physics, and enzymology, illustrating the diverse areas in which the program is able to develop talent.

The value to the graduate student is a disciplined educational program with quality mentorship from mathematicians and computer scientists, as well as career opportunities at DOE national laboratories and the industrial or educational institutions where their skills are appreciated and sought. Currently 37 fellows are enrolled at 22 major universities. There are 11 practicum sites, including large, multipurpose laboratories with active DOE projects.

Woods Hole Geophysical Fluid Dynamics Summer Program[11]

For nearly four decades, the Geophysical Fluid Dynamics Summer Program, located at the Woods Hole Oceanographic Institution, has brought together mathematicians and fluid dynamicists from diverse areas of frontier research and from many institutions. Each 10-week summer program focuses on a particular research theme. The course begins with an introductory formal lecture series, followed by a summer-length sequence of research seminars. Graduate students admitted to the course are required to initiate research projects under the close guidance of the staff on topics connected to the particular theme of the summer course. The program is guided by a core of senior faculty, whose presence ensures long-term continuity. This core is

---

[11]Further information on this program can be found at <http://www.whoi.edu/education/dept/#GFD>.

supplemented by a flow of long- and short-term visitors who contribute special insights to each summer's theme.

The program has been influential in establishing the value of a mathematical foundation for research in geophysical fluid dynamics. It has a history of involving both mathematicians and fluid dyamicists. This long-term involvement has broken down barriers between the two areas, as vocabulary and perspectives are shared and mutually appreciated, and has been vital to the program's success. At the same time the flow of students through the program has populated the field of geophysical fluid dynamics with a cohort that has learned the benefits of interaction between mathematicians and scientists. The program has also served to educate senior faculty.

The summer course has several important components that have led to successful math-science collaboration, foremost among them the graduate student component. The educational aspect of the program is the key to its long-term influence on both mathematical science and other sciences. Long-term support for this initiative and its summer-long duration have allowed communication barriers between the disciplines to be overcome. The infrastructural support provided by the Woods Hole Oceanographic Institution enables the course, provides the proximity that fosters research interactions, and allows participants to readily interact with Woods Hole scientists, all of which magnify the benefits of the program.

SUCCESS FACTORS

The programs cited above by the committee are by far not the only programs, nor have they been scientifically chosen. Nonetheless they point to a number of factors that seem to be a regular feature of successful cross-disciplinary research. The committee believes that these factors can be usefully applied to other efforts to create linkages led by scientists or mathematicians:[12]

- *The programs bring collaborators together for an extended period of time or for repeated interactions over a long period of time.* The Courant Institute, the Institute for Mathematics and Its Applications (IMA), the Mathematical Sciences Research Institute (MSRI), the National Center for Supercomputing Applications (NCSA), the University of Michigan Mathematics Department, and the Woods Hole programs all provide researchers with enough time together, which is a fundamental part of their strategy. These periods of time vary in length, and some positions are resident and some are not. Some programs allow researchers to choose their length of stay (a full year or a half year, for example). While the programs often sponsor seminars and symposia, thereby reaching a larger audience and providing short, one-time contact between researchers from various disciplines, clearly it is from those researchers who spend prolonged periods that the strongest and most fruitful research interactions spring.
- *Many programs have a formal educational component.* The Carnegie Mellon, Courant, IMA, Spelman, University of Arizona, University of Chicago, University of Michigan, and Woods Hole programs all have a formal curriculum. This has a twofold impact: it trains a new generation of researchers in cross-disciplinary thinking and it challenges the established

---

[12]It should be noted that while most of the programs are based in mathematics departments or institutes, the mathematical community does not bear primary responsibility for fostering such linkages.

researchers who teach the courses to broaden their own ideas. Virtually the entire community of U.S. geophysical fluid dynamics researchers can trace its roots to researchers educated at the Woods Hole summer program, testifying to the long-term impact such programs can have.

- *Strong leadership is needed to set up a successful program.* The Courant, IMA, Spelman, University of Chicago, and University of Michigan programs all show the importance of a visionary leader in establishing a successful cross-disciplinary research program, and the NCSA program shows the importance of continued leadership in sustaining excellence. Sometimes this leadership came from a small group of investigators, sometimes from a program officer at a funding agency or from the management of a research center, and sometimes from a department head. These people were able to appreciate the importance of creating a cross-disciplinary interface and to communicate it to all the players involved, giving them a rationale to expend the resources necessary to achieve success.

- *Mentoring lowers the barriers to successful cross-disciplinary research.* This is a component of all the programs but is seen most clearly in the Carnegie Mellon, Spelman, University of Arizona, and Woods Hole programs. Active mentoring of students helps assure that they receive appropriate training and identifies employment opportunities. The presence of good role models on the faculty makes it easier for students to consider cross-disciplinary work. Something as simple as faculty attitude, as reflected in attendance at cross-disciplinary seminars and willingness to teach cross-disciplinary courses, delivers a strong message to students about the importance of cross-disciplinary training and breaks down the psychological barriers to crossing the lines between disciplines.

REFERENCES

American Mathematical Society (AMS), Task Force on Excellence. 1999. Towards Excellence: Leading a Mathematics Department into the 21st Century. Available at <http://www.ams.org/towardsexcellence>.

Colton, T., and P. Armitage, eds. 1998. Encyclopedia of Biostatistics. New York: John Wiley & Sons.

Deans, S.R. 1983. The Radon Transform and Some of Its Applications. New York: John Wiley & Sons.

Gul, F. 1997. A Nobel prize for game theorists: The contributions of Harsanyi, Nash and Selten. Journal of Economic Perspectives 11:159-174.

Herman, G.T., ed. 1979. Image Reconstruction from Projections. New York: Springer-Verlag.

Johnson, N., and S. Kotz, eds. 1982. The Encyclopedia of Statistical Science. New York: John Wiley & Sons.

Leonard, R.J. 1995. From parlor games to social science: von Neumann, Morgenstern and the creation of game theory. Journal of Economic Literature 33(2):730-761.

Lindzen, R.S., E.N. Lorenz, and G.W. Platzman, eds. 1980. The Atmosphere—a Challenge: The Science of Jule Gregory Charney. Boston, Mass.: American Meteorological Society.

Nasar, S. 1998. A Beautiful Mind. New York: Simon and Schuster.

National Research Council (NRC). 1994. Research Directorate Programs in the United States: Continuity and Change. Washington, D.C.: National Academy Press.

Natterer, F. 1986. The Mathematics of Computerized Tomography. New York: John Wiley & Sons, p. 289.

Platzman, G.W. 1979. The ENIAC computations of 1950: Gateway to numerical weather prediction. Bulletin of the American Meteorological Society 60:302-312.

Stigler, S. 1986. The History of Statistics. Cambridge, Mass.: Belknap Press.

3

Previous Efforts and Studies Relevant to Math/Science Linkages

The committee is not the first to consider the need for cross-disciplinary linkages and how to forge them. Many other efforts have focused on or considered the contributions that math-science linkages can make to the development and timely advance of frontier areas of research (Appendix C). It is difficult, however, to demonstrate quantitatively the need for further cross-disciplinary research links. Data on graduation rates, research funding, and other quantitative measures are routinely collected for the various individual disciplines but would be difficult to collect for cross-disciplinary research, as even the definition of "cross-disciplinary" is a topic of debate among the various research communities. Even if such data were available, they could not be used to project how many more cross-disciplinary researchers are needed and should be trained every year. While this exercise has been done for disciplines with well-defined and agreed-upon goals (see, for example, NRC, 1984), no such consensus exists on scientific cross-disciplinary research. Accordingly, the evidence suggesting cross-disciplinary linkages should be strengthened is qualitative rather than quantitative: a growing body of organizations, individuals, committees, and other entities have indeed recognized the importance of cross-disciplinary research to advancing the frontiers of their fields, and a number of programs have been set up to begin to address this need (Appendix D).

National interest in cross-disciplinary research peaked sharply after World War II and began rising again in the mid-1980s. In 1986, the Chairman of the House Committee on Science and Technology emphasized that interdisciplinary solutions would be required for the major scientific and technological problems of the world, claiming that the "mega-problems [are those] that cannot be neatly pigeonholed into any one of the traditional academic disciplines" (U.S. Congress, House, 1986). This new interest was presumably in response to the increasing amount of data being generated and stored by scientists, the richness of the systems they investigate, and a perceived disconnection between scientists and mathematicians (Sigma Xi, 1988).

This growing interest in linking mathematics and science is most easily observed in reports generated by the mathematical community. The so-called David I report (NRC, 1984) identified a significant decline in funding for mathematics despite its increasing relevance to science, technology, and society. It gave many examples of important applications of mathematics. Six years later, the "David II" report identified significant opportunities for mathematics research, including cross-disciplinary collaborations (NRC, 1990). The David II report also recommended that academic mathematical science departments give more recognition to faculty engaging in cross-disciplinary collaborations. In 1994, the Joint Policy Board for Mathematics encouraged the mathematical sciences community to develop a rewards system that does not distinguish between core and applied mathematics and that values a number of activities, including teaching, outreach activities, and cross-disciplinary pursuits (AMS, 1994). Other recent reports have warned that U.S. mathematics should pay serious attention to its interactions with other disciplines in order to remain preeminent in the field (NRC, 1997; AMS, 1999). The *Report of the Senior Assessment Panel of the International Assessment of the U.S. Mathematical Sciences* (NSF, 1998) bases its evaluations in part on the rate at which new mathematics is utilized by other disciplines. The report argues that by retreating from multidisciplinary involvement, mathematics misses opportunities to be enriched by the ideas and challenges of other disciplines. The other disciplines suffer from a loss of expertise and easy access to the vast knowledge base being developed by the mathematical sciences and from the overly specialized mathematical languages and tools that inhibit communication with other disciplines. According to this report, "strengthening the connections between the creators and the users of mathematics, while maintaining historical proficiency in pure mathematics, is the most important opportunity now open for the National Science Foundation in support of that field."

Reports from the scientific community likewise acknowledge the benefits of research at the interface between science and mathematical sciences and of providing students, postdoctoral fellows, and other researchers with more training in the mathematical sciences. For example, a growing interest in math-related research and training is evident from reports and program initiatives coming from the life sciences and biomedical research community. The David II report, mentioned above (NRC, 1990), suggested ways of better reviewing and otherwise promoting collaborations between life and medical scientists, on the one hand, and physical scientists (including mathematical sciences) and engineers, on the other. In 1992 and 1996, the National Science Foundation published reports detailing the historical relationship between mathematical sciences and biology and suggesting areas of emerging opportunity at the math-biology interface (NSF, 1992 and 1996). More recently, four foundations supportive of biomedical research convened a workshop to discuss the impact of market forces on health care research in the United States and the ways in which private foundations could best "meet the challenging needs of research and advance in the field" (American Cancer Society, Burroughs Wellcome Fund, Howard Hughes Medical Institute, and the Pew Charitable Trusts, 1998). The resulting report lists seven emerging themes identified at the workshop. One of these acknowledges the potentially high payoff associated with crosscutting research in biology and the mathematical sciences. Expressing concern about the financial crises faced by academic health centers (AHCs), the report claims that general support of multidisciplinary research can be critical to the long-term success of an AHC program because it facilitates breakthrough research; maximizes the attractiveness of an institution to graduate students, postdoctoral trainees, and outstanding faculty; brings the research community together; and increases the likelihood of facilitating linkages with the commercial sector. Other disciplines have similarly expressed the need to promote research linkages with the mathematical sciences community.

Appendix C briefly describes these and other studies that considered how to establish better research linkages as well as some actions taken to remedy the situation. The chronology demonstrates a growing recognition of the importance of interdisciplinary research to the health and advancement of U.S. science and technology generally and the necessity of cross-disciplinary linkages in maintaining and increasing the health of the mathematical sciences in particular. Several studies recognize the special role mathematics has to play in the advancement of the research infrastructure as sciences that were traditionally descriptive become more and more quantitative.

The studies repeatedly identify certain factors that must be addressed to foster research linkages. Many recommend increased interdisciplinary training at all levels and the integration of interdisciplinary problems and experiences into disciplinary curriculums, and many discuss cultural issues such as the need for open communications and collegiality between disciplines. The need to reform current institutional structures or build new ones is noted again and again. These structures include incentives such as promotion and tenure criteria, which generally need modification if they are to fully value cross-disciplinary research, and disciplinary departments that can introduce new structures and means to promote interaction with colleagues from other departments. Funding agency structures, which work well for disciplinary research proposals but often lack mechanisms for adequately reviewing cross-disciplinary work, also need reform. Finally, the need for greater resources, both to strengthen the health of mathematics and to increase the number of cross-disciplinary research linkages, is repeatedly noted.

This body of work represents the continuing effort of a number of research communities to encourage, evaluate, and prioritize efforts to promote cross-disciplinary research linkages. As a whole it arrives at the same conclusion—that cross-disciplinary research linkages are crucial to many important advances in science and technology and to health of the research enterprise itself. And yet, no action plan for moving the U.S. research infrastructure forward has emerged. The committee was left to ask itself what more it could say in the face of such a large body of earlier work. It concluded that it could offer some detailed, specific recommendations for programs that would enhance interactions at the math-science interface. In addition, it felt that limited efforts, such as those represented by its own work and that of the committees that had preceded it, would not be sufficient to move the research community and funding organizations forward or to ensure that math-science interfaces were enhanced. Instead, the committee felt a more continuous effort and oversight was required. Chapter 4 discusses some steps the committee feels can be taken to strengthen cross-disciplinary research in the United States.

REFERENCES

American Cancer Society, Burroughs Wellcome Fund, Howard Hughes Medical Institute, and the Pew Charitable Trusts. 1998. Strengthening Health Research in America: Philanthropy's Role. Available at <http://www.pewtrusts.com/pubs/publications.cfm>.

American Mathematical Society (AMS). 1994. Recognition and Rewards in the Mathematical Sciences. Washington, D.C.: AMS.

American Mathematical Society (AMS), Task Force on Excellence. 1999. Towards Excellence: Leading a Mathematics Department into the 21st Century. Available at <http://www.ams.org/towardsexcellence>.

National Research Council (NRC). 1984. Renewing U.S. Mathematics: Critical Resource for the Future. Washington, D.C.: National Academy Press.

National Research Council (NRC). 1990. Renewing U.S. Mathematics: A Plan for the 1990s. Washington, D.C.: National Academy Press.

National Research Council (NRC). 1997. Preserving Strength While Meeting Challenges: Summary Report of a Workshop on Actions for the Mathematical Sciences. Washington, D.C.: National Academy Press.

National Science Foundation (NSF). 1992. Mathematics and Biology: The Interface—Challenges and Opportunities. PUB-701. Berkeley, Calif.: Lawrence Berkeley Laboratory.

National Science Foundation (NSF). 1996. Modeling Biological Systems. A Workshop. Available at <http://www.nsf.gov/bio/pubs/mobs/stmobs.htm>.

National Science Foundation (NSF). 1998. Report of the Senior Assessment Panel of the International Assessment of the U.S. Mathematical Sciences. NSF9895. Arlington, Va.: National Science Foundation.

Sigma Xi. 1988. Removing the Boundaries: Perspectives on Cross-Disciplinary Research. Research Triangle Park, N.C.: Sigma Xi Publications.

U.S. Congress, House of Representatives. 1986. American Science and Science Policy Issues: Chairman's Report to the Committee on Science and Technology, U.S. House of Representatives. Washington, D.C.: Government Printing Office.

4

Recommendations

The first three chapters identified some of the enabling factors as well as some of the obstacles to successful interactions between mathematics and the sciences. The committee hopes the recommendations in this chapter will enhance existing linkages, stimulate new interactions, and help to overcome some of the obstacles. It expects that the implementation of these recommendations will increase the flow of interesting and challenging ideas between the sciences and mathematics and improve communication between the two groups.

FUNDING DIRECTIONS

The committee recommends that additional funding be allocated to initiatives that will strengthen existing linkages between the mathematical sciences and other sciences and that will build new linkages.

New resources are required if the nation is to realize the enormous benefits of cross-fertilization. The committee believes the responsibility lies equally with the practitioners of mathematical sciences and the sciences. Some disciplines can benefit from a shift in some of their resources from strictly disciplinary to cross-disciplinary activities. Others are too close to the critical mass of necessary resources to benefit from such shifts. Since productive cross-disciplinary research cannot be built on a base of foundering disciplines, some disciplines will need additional monies to be capable of developing effective cross-disciplinary ties. The health of the basic sciences is a prerequisite for the advance of cross-disciplinary research, so funding for cross-disciplinary research must not compromise support for basic disciplinary research or individual investigators. The committee expects that successful linkages will expand the horizons of disciplinary research in many directions in both the mathematical sciences and other sciences. It offers here specific recommendations for programs that would foster cross-disciplinary research, along with an estimate of the cost based on the committee members' experience in running similar programs.

Enhancing Multidisciplinary Activities

Increase the number of specialized summer institutes, each sustained for at least 5 years, organized around a core of committed senior scientists and mathematical scientists, and aimed at fostering linkages between the sciences and mathematical sciences.

The Geophysical Fluid Dynamics Program at Woods Hole, described in Chapter 2, could serve as a model for introducing senior scientists, graduate students, and postdoctoral fellows to cross-disciplinary research and for sustaining their interest in and commitment to such research. The success of that program lies in the long-term continuity assured by a core group of senior faculty, combined with its educational aspect—the training of graduate students during an intense, summer-long program. The committee recommends that other such institutes be created that are devoted to bridging the mathematical sciences and other sciences and having the same elements of long-term continuity in funding and education. Such institutes would foster the prolonged interactions necessary to establish meaningful cross-disciplinary collaborations, provide researchers opportunities to network with colleagues from other disciplines, and help a core group of researchers establish a sufficient understanding of each other's disciplines to recognize promising research opportunities at the disciplines' interface.

Proposals for such institutes should be subject to vigorous competition and peer review. Based on the committee's experience, an institute could be established and continued for a cost of about $2 million per year. As these institutes mature, costs may decrease—the cost of the well-established Geophysical Fluid Dynamics Program, for example, is currently about $150,000 per year.

Encourage existing large-scale programs (such as research institutes, NSF science and technology centers (STCs)) or Department of Defense-University Research Initiatives (multidisciplinary university research initiatives) to develop targeted initiatives when promising ideas need new linkages between science and the mathematical sciences.

Existing large-scale programs provide an infrastructure that can be leveraged to build new math-science linkages. Agencies can encourage centers to explore new research topics at little additional cost, as the centers already have the framework in place to run workshops and pursue novel directions. For example, the NSF-STC for Discrete Mathematics and Theoretical Computer Science (DIMACS) sponsored workshops on mathematical support for molecular biology from 1994 to 1996. These workshops attracted many biologists, chemists, computer scientists, and mathematical scientists. Also, the Aspen Center for Physics, the Institute for Advanced Study, the Institute for Theoretical Physics, and the Mathematical Sciences Research Institute all sponsored workshops and programs that supported the exciting connections between high energy physics and modern mathematics. These activities provide unique opportunities for networking between disciplines and for educating researchers about problems outside their own disciplines.

Fellowship Programs to Sustain Research Scientists Pursuing Promising Multidisciplinary Ideas

Support long-term crossover visitor and sabbatical programs by which mathematical scientists could visit science departments and laboratories and scientists could visit mathematical science departments.

Sabbatical programs would provide opportunities for researchers to be colocated with colleagues from other disciplines for long enough times to establish meaningful collaborations. The programs could last for 6 months or a year or they could involve participation in specific courses targeted at math-science collaboration for shorter periods. These cross-disciplinary sabbaticals should be highly competitive and endowed with particular prestige, both to encourage researchers to participate and to make them a plus in participants' promotion files. It would be useful to distinguish such sabbatical awards with an appealing title, such as the von Neumann, Wiener, or Fisher sabbatical visitor. The committee estimates the cost for each exchange at $70,000.

Establish cross-disciplinary research grants with sustained funding for at least 5 years to allow young investigators to collaborate across the mathematical sciences and the sciences.

Such sustained support would help talented young researchers with vision pursue cross-disciplinary research, which by its nature poses greater career risks than traditional disciplinary research and often takes longer to show substantial results. The committee estimates a reasonable level of support to be $250,000 per investigator for the 5 years.

Establish cross-disciplinary, postdoctoral-plus fellowships, highly competitive grants for postdoctoral research that would provide continuing research support once the recipient obtains a tenure-track academic position.

Similar to the two programs described above, these fellowships would consist of a postdoctoral fellowship followed by a research grant for the first 2 tenure-track years. The exceptional young researcher could then begin an academic career with full momentum and move ahead on whatever ideas come from the postdoctoral years. These grants would address the difficulty faced in securing funding for research that falls between traditional disciplines and would provide an additional year of research time to offset the additional time often required to establish good cross-disciplinary research. Such funding should not preclude the possibility of teaching, as is the common practice among postdoctoral fellows in mathematical sciences. Cross-disciplinary teaching might even be encouraged. Such a proposal might be funded at a level of $325,000 per investigator for a 4-year period.

CROSS-DISCIPLINARY INTERACTIONS

The committee recommends that academic institutions take responsibility for implementing vigorous cooperative programs between the sciences and the mathematical sciences.

Universities with vision are developing new programs that cut across traditional departmental and college boundaries. Such programs are being recognized as a significant component of a university's research and educational missions. The mathematical sciences, because their role is central to all the sciences, are well placed to participate fully in the dramatic changes taking place. Indeed, a recent AMS report (AMS, 1999) suggests that forming good ties to the university's mission and to other academic departments is key to the success of a mathematics department and discusses in depth several departments that have used this approach to strengthen their programs. Astute mathematical science departments and cooperating, sympathetic senior administrators can create a hospitable environment that encourages research scientists to pursue exciting cross-disciplinary activities and can develop the curricula needed to give students cross-disciplinary skills. Academic programs in sciences such as engineering and biology can encourage cross-disciplinary educational programs as departments modify curricula to meet the evolving needs of their disciplines. Leadership at the university and departmental levels can do much to build and sustain cross-disciplinary programs and, in so doing, to enhance the U.S. research enterprise.

No cross-disciplinary program will succeed unless the institution demonstrates that it values and is committed to such efforts. In the end, no other attempts to convey that message will succeed unless the reward structure for researchers is consistent with the message.

Departments and colleges should develop effective criteria for the evaluation of cross-disciplinary research, ensuring that the university's promotion, tenure, and reward mechanisms fully recognize quality cross-disciplinary research.

It is imperative to improve evaluation criteria for investigators pursuing cross-disciplinary research in order to demonstrate its value to the university. Particularly helpful would be protocols for the inclusion of extradisciplinary expert testimonial in departmental and college promotion and tenure committees. There are institutions in which interdisciplinary work is naturally accepted and effectively handled in promotion and tenure. One example is medical schools, where the basic science and clinical science departments follow procedures of faculty vote and approval similar to those of the basic science departments at universities and where the tenure committees include faculty from other disciplines. Universities with vision must recognize that the implementation of such criteria is critical for the health of cross-disciplinary endeavors. This is one of the most important steps that can be taken to change attitudes and cultures that view cross-disciplinary research as second-rate or less important than disciplinary work.

OVERSIGHT

The committee recommends that a new standing committee be established with a long-term focus on improving the linkages between the mathematical sciences and other sciences in both academia and industry.

In reviewing earlier studies of the linkages between the mathematical sciences and other sciences, the committee was struck by the number of reports that had had relatively little impact, despite their good ideas. The committee felt one reason for this was the lack of an effective follow-through mechanism. An ongoing group with special responsibility for advocating cross-disciplinary projects between the mathematical sciences and other sciences could nurture linkages. The group could monitor and evaluate the various initiatives to foster linkages and communicate their evaluations to the community. The committee would be a source of information on opportunities for research linkages.

A standing committee devoted to maintaining the health of linkages between science and mathematical sciences would be asked to do the following:

- Take a proactive role in alerting funding agencies and policy makers to promising new cross-disciplinary directions;
- Promote, compile, and publicize opportunities for cross-disciplinary work;
- Advise administrators setting up cross-disciplinary programs and curricula involving the mathematical sciences and acquaint them with successful models;
- Increase the visibility of cross-disciplinary mathematical research to the scientific community by disseminating the results of such research through workshops, conferences, and publications;
- Monitor and identify effective mechanisms by which academic institutions and funding agencies can assess cross-disciplinary research proposals; and
- Report regularly on the status of linkages between the mathematical sciences and other sciences.

The standing committee should be distinct from current disciplinary bodies so it can maintain a focus on cross-disciplinary education and research. It should be distinct from existing groups representing specific disciplines to avoid diluting the missions of those bodies and also to avoid the perception that it is advocating one discipline over another. Nonetheless it would be critical for the new committee to have representation from existing bodies in order to have a strong link to the communities that would carry out its suggestions.

The committee could be established on a 5-year trial basis, with an evaluation of its effectiveness at 2 years to allow adjusting its efforts. An evaluation after 5 years would determine whether it should be continued. The committee's activities could be funded by agencies and foundations interested in fostering math-science research linkages.

The standing committee would advise funding agencies on how they might better evaluate cross-disciplinary research.

The committee recognizes and applauds the efforts already under way by federal agencies to increase math-science linkages. For example, NSF's Office of Multidisciplinary Activities (OMA) supports particularly novel, challenging, or complex multidisciplinary research projects that do not fit well into the existing program structure or whose realization might otherwise be hampered by existing institutional and procedural barriers. Its new Integrative Graduate Education and Research Training (IGERT) program is evidence of NSF's commitment to multidisciplinary approaches to graduate education. NIH is actively involved in new funding and research at the interface of mathematical sciences and science. Its recently announced Fellowships in Quantitative Biology program, for example, encourages highly qualified individuals with doctoral training in the traditional quantitative disciplines to obtain training in biological areas congruent to the mission of the National Institute of General Medical Sciences (NIGMS). DOE is currently advancing a major initiative in the application of computational science to science and engineering problems. DARPA, DOE, and ONR have a long history of funding cross-disciplinary projects related to their missions. (Some current efforts to increase the linkages between the mathematical sciences and other sciences are described in Appendix C.)

However, because the funding agencies were established when scientific disciplines were strictly separate, their organizational structure can discourage promising multidisciplinary research. As an example, cross-disciplinary proposals are sent to different panels with different goals and cultures and so are evaluated differently, making it seem as if it is more difficult to get them funded. Some mechanism needs to be devised so agencies can get an objective scientific evaluation of cross-disciplinary proposals. As another example, the application of mathematics to the biomedical sciences is a frontier research area vital to the NIH mission. Organizing a permanent study section composed of scientists from different disciplines, including the mathematical sciences, might be one inexpensive way for NIH to ensure the proper consideration of highly mathematical proposals. In any event, a committee could advise federal agencies on how to evaluate cross-disciplinary proposals by describing the methods of organizations that have successfully done so.

The new committee could advise private foundations and philanthropies on effective methods for supporting promising cross-disciplinary research.

Private foundations such as the Pew Charitable Trust, the Keck Foundation, Burroughs-Wellcome, Howard Hughes, and others foster collaborative research through their support for training programs and for faculty research, training, and in some cases, new hires. Since research at the interface of the mathematical sciences and other sciences is playing an increasingly important role in all areas of research, especially biomedical research, considerable leveraging can be anticipated for investments that foster such research. Private foundations can design and fund the sort of innovative institutes, fellowships, and research positions that the committee believes will effectively strengthen and increase collaborative research efforts.

The new committee could help identify curriculum changes that would enable more and better linkages.

Many scientific disciplines are recognizing the increasing importance of mathematics to the success of their fields and are adjusting their educational programs accordingly.

Mathematical science departments are also increasingly recognizing the importance of exposing their students to applications in the sciences and are changing their programs accordingly. A standing committee could convene and facilitate dialogue on curricular modifications at both the undergraduate and graduate level, enabling an ongoing examination of efforts undertaken throughout the country and disseminating information on successful reforms.

The new committee could inform professional societies about multidisciplinary research opportunities and related educational opportunities.

Professional societies can promote good cross-disciplinary research. By combining forces, societies representing different disciplines can disseminate information about cross-disciplinary research and funding opportunities through mechanisms such as special sessions at their annual meetings, articles in membership journals, symposia, and newsletters. A committee concerned specifically with cross-disciplinary research would help the many professional societies to collaborate in the pursuit of these goals. It could also help these societies discuss and disseminate ideas and recommendations for curriculum changes and professional development pertinent to cross-disciplinary research. It would cite examples of departments that have successfully forged substantial cross-disciplinary linkages and would highlight the mechanisms used to achieve that success.

There are several ways to organize and administer such a committee, but certain criteria for its composition are critical. It must contain members who would be perceived as respected representatives of relevant disciplines in mathematical and other sciences. It is important for the membership to have the capacity to relate the committee's deliberations and findings to other standing committees of the various professional societies, federal agencies, and the National Academies; to that end it would be ideal to draw a portion of the membership from such standing bodies. Because of the important influence educational programs can have on building the math-science interface, the membership must also include individuals viewed as influential educators in their disciplines. Founding such a group would improve the likelihood of follow-up to the findings of this committee.

REFERENCE

American Mathematical Society (AMS), Task Force on Excellence. 1999. Towards Excellence: Leading a Mathematics Department in the 21st Century. Available at <http://www.ams.org/towardsexcellence>.

APPENDIXES

A

Ten Case Studies of Math/Science Interactions

1. Modeling Weather Systems Using Weakly Nonlinear, Unstable Baroclinic Waves
Joseph Pedlosky
Department of Physical Geography
Woods Hole Oceanographic Institution

The pioneering work of Jule Charney and Eric Eady in the late 1940s showed how the development of large-scale disturbances in atmospheric flow patterns that were associated with emerging weather systems could be explained as a natural instability of the largely westerly, zonal (west to east) winds in Earth's atmosphere. Using a linear, small-amplitude perturbation theory, they were able to explain the basic energy source of the weather wave in terms of the ability of the perturbations to tap into the potential energy distribution in the initial, basic flow on which the disturbances grew and from which they efficiently extracted energy. The linear theory gave reasonably good predictions for the spatial scale of the instabilities and their initial rate of growth. What was lacking in these models was an explanation of what limited the growth of the disturbances as they drained energy from the basic flow, at what amplitude saturation would occur, and what dynamical state would follow that saturation.

Some years ago, I became interested in developing the theory for weakly nonlinear, unstable baroclinic waves as models of weather systems in the atmosphere and eddies in the ocean. The weakly nonlinear theories existing at the time followed the work of Stuart and Watson (for the Orr-Sommerfeld problem) or the Malkus and Veronis approach (for convection). In each of these cases the threshold for instability is determined by a balance of energy release by the unstable mode against internal energy dissipation. That means that the marginally unstable wave already has to have a structure that extracts energy from the mean state. Therefore, a relatively straightforward perturbation expansion around that starting point could give the equation for the evolution of the amplitude of the disturbance in the form suggested by Landau, i.e.

$$dA/dt = sA - NA^3$$

where s is the linear growth rate and N is the fruit of the perturbation calculation (the Landau constant).

However, in the geophysical problem the threshold for instability is determined by overcoming inviscid (adiabatic) constraints associated with what is called potential vorticity conservation. That means the marginal wave (around which a perturbation expansion is started) neither dissipates nor extracts energy. In the absence of an energy-releasing structure in the marginal wave, it is impossible to calculate the effect of the perturbation on the mean and hence to get at the Landau constant.

Fortunately for me, a colleague at Chicago was deep at work on the problem of resonant interactions of capillary waves, and it occurred to me that the method of multiple time scales he was using would allow me to calculate the phase shifts in the wave required for energy extraction in an implicit way that would subsequently be determined by the evolution itself. This led to a new form of the evolution equation, which has a lot in common with the following equation:

$$d^2A/dt^2 + rdA/dt = -s^2A + NA|A|^2$$

where r is a measure of the very weak dissipation in the evolving wave. Actually the geophysical problem is a bit more complicated but not much more so (the last term is really a spatial integral over a function that satisfies a partial differential equation that is first order in time but whose amplitude is $O(|A|^2)$. Hence the full system is third order in time.

By another piece of good luck I spent the following summer visiting the Geophysical Fluid Dynamics (GFD) program at Woods Hole, which has always attracted a fair number of mathematicians interested in physical problems, for example, Lou Howard, Joe Keller, and Ed Spiegel (who is at the intersection of math and astrophysics). One of them recognized that the equation I had derived might be put in the form of the Lorenz equations (again a slight oversimplification of the situation—the partial differential equation again). This turned out to be true. It was the first time that I know of that the Lorenz equations were derived as a systematic asymptotic approximation to weakly nonlinear theory instead of an arbitrary, low-order, truncation of a Fourier expansion of a strongly nonlinear problem (where it usually fails as a solution to the original problem when the nonlinearity is important enough to be interesting—this is the part of the Lorenz theory of convection nobody mentions).

Another mathematician happened to be visiting the Massachusetts Institute of Technology and asked me after a seminar I gave there whether my results fit the Feigenbaum scenario for the map problem. My immediate reaction was, "Who is Feigenbaum?" With some further help I was able to take advantage of all of the work done by Feigenbaum and others. Indeed, perhaps not surprisingly, the map theory was a good conceptual organizer for the results of the baroclinic instability problem.

I think that body of work, which evolved further with the help of lots of other people (among them John Hart, Patrice Klein, and Arthur Loesch), was fundamentally enriched at the start by chance interactions with people who knew a lot more math than I do. That happened because certain venues (e.g., the Woods Hole GFD program) had patiently established an atmosphere where unprogrammed interactions of that type could take place if only the participants (1) met frequently enough to know who to go to for help, (2) had established some common language, and (3) shared an interest in the boundary regions between math and (in this case) geoscience.

2. Mixing in the Oceans and Chaos Theory
Larry Pratt
Department of Physical Oceanography
Woods Hole Oceanographic Institution

One of the central problems in physical oceanography is understanding the transport and mixing of chemical and physical properties such as heat, salinity, ice, and nutrients. These quantities generally are carried from one place to the next by relatively narrow and intense current systems such as the Gulf Stream and Kuroshio. However, there also is a great deal of mixing that takes place across the edges of these currents, causing a slower transport of properties into the surrounding fluid. Understanding how this mixing occurs has been a challenge for many years.

A relatively new approach to this problem has recently been developed within the branch of applied mathematics known as "dynamical systems," a branch concerned, in part, with the science of chaos. Applied mathematician Chris Jones of Brown University and I have been working together over the past 5 years to develop the new approach and use it to gain new insights about ocean mixing.

We first met in 1992, when Chris was visiting the Woods Hole Oceanographic Institution (WHOI) with his colleague from Caltech, Steve Wiggins. The two were working on a project supported by the Office of Naval Research (ONR), and their program manager, Reza Malek-Madani, had encouraged them to find out more about physical oceanography.

Hence their visit to WHOI. They spent the day talking to WHOI staff, including me, about research problems involving ocean waves, currents, and mixing. This was unfamiliar territory to Chris, so we never would have run into one another had Malek-Madani not served as a matchmaker.

Over the next months, Chris and his mathematician colleagues continued their crash course in physical oceanography by meeting researchers from different institutions. With funds from ONR, Chris was able to arrange a 2-day workshop in 1993 to get all the mathematicians and oceanographers together at a beautiful inn near Little Compton, Rhode Island. Each researcher presented his interests to the gathering and there was plenty of time for casual discussion.

The workshop helped participants to begin crossing into each others' cultures. About a year later, Chris and I formulated the conceptual basis for what became a long-term collaboration.

Chris recalls the meeting as "a critical event in the process while not seeming so at the time." For one thing, his initial discussions with me were far away from the transport issues that became the basis of our subsequent collaboration. And it was other researchers, including Wiggins, who were presenting material on transport. Yet Chris and I ended up collaborating on the transport issues.

This is good example of the long "spin-up" time often required for such work to get under way. The initial education stage cannot be rushed.

In 1993, the idea of a collaboration became real. Since then, a core group of four or five oceanographers, an equal number of applied mathematicians, and various students and postdoctoral scholars has continued to work together. Workshops at Caltech, Woods Hole, and the University of Minnesota have been a primary means of information exchange. The workshops do in fact take work. For one thing, most of the oceanographers in the group approach mathematics somewhat as a hobby. So it takes patient and articulate colleagues, including Chris and Steve Wiggins, to bring the ocean scientists up to speed on the relevant developments in mathematics and chaos theory.

Our workshops are occasionally attended by very well known researchers, but their participation has not catalyzed any collaborations that I know of. There is an important sociological point to be made here: it is not enough for people to be smart, they also have to be able to get along.

My work with Chris has centered on applying dynamical systems techniques to understand mixing and transport processes in the ocean. There is nothing new about applying dynamical systems methodology to the understanding of mixing, but that had not previously been done using a particular technique, lobe dynamics. The method involves calculating both stable and unstable manifolds in models of meandering ocean currents and recirculations.

That particular course of action began about a year after the Little Compton workshop in 1992. I had been struggling to understand a particular mixing problem that arises in connection with the Gulf Stream. Chris suggested using lobe dynamics, under the hunch that it might be better able to describe the Gulf Stream mixing phenomena than other techniques. The method allows quantification of a new transport process, and Chris suspected it also might be applicable for mathematically describing a variety of meandering ocean currents.

He was right, but applying lobe dynamics to the Gulf Stream problem would take some adjustments. For one thing, our Gulf Stream models remained incomplete (lacking, for example, analytically specified vector fields and infinite time). What might have seemed like a headache at first, ended up by catalyzing a benefit. This led to further developments, such as generalized definitions of manifolds over finite time. So while oceanographers were getting new tools to do their science, mathematicians ended up with new directions to take their math.

Postdoctoral scholars, assistant professors, and graduate students also have played key roles in the ongoing collaboration. They're also human carriers of any valuable cultural hybridization that comes out of such collaborations. That's why it is important for these participants to receive professional recognition for their work.

The recognition issue poses special challenges in these kinds of collaborations because the cross-disciplinary content of the resulting papers does not fit neatly within traditional disciplinary boundaries. Young physical oceanographers publishing in *Physica D,* for example, will be read mostly by applied mathematicians but not by the oceanographers in their own field. This is more than just a theoretical hardship. It has been a real problem for one of the younger team members working with us.

The professional culture of mathematics has been more accommodating when it comes to credit due. Applied mathematicians do in fact receive credit for publishing in, say, the *Journal of Physical Oceanography,* since the work is, as a result, perceived by their colleagues as relevant.

The general problem of reforming the criteria for professional recognition to accommodate interdisciplinary collaborations remains unsolved. The strongest resistance I have encountered in getting our work recognized has been within my own field, physical oceanography. The field has a large contingent of observationalists, who must devote vast amounts of time to planning scientific cruises, designing instruments, and processing data. For them, mathematics is not even a hobby, and they view mathematicians who work on oceanography with suspicion. It is sometimes difficult to get this important part of the field to learn enough about dynamical systems to appreciate the value of our work. The challenge, however, makes the eventual acceptance of our work that much more satisfying.

There are no prescriptions or templates for interdisciplinary collaborations. They are adventures in problem solving. As Chris points out, there is an extra sociological dimension involved in bridging disciplines. As in any successful partnership, both sides need to be equally committed to the collective goal. A lopsided distribution of interest can poison the effort.

The culture of mathematicians may be able accommodate the flexibility needed for successful collaborations between its own members and the other scientific disciplines. But Chris says that mathematicians willing to work on two levels at the same time will be crucial players. First, they should work on problems of short-term interest to the collaborating applied scientist. That will help to win the attention and respect of the scientists by demonstrating how unfamiliar mathematical techniques can help them do more. Secondly, the mathematician should glean from the work some deeper mathematical problems that promise to further develop mathematical theory and to suggest future directions, enriching the profession.

The collaboration that developed between me, a physical oceanographer, and Chris, an applied mathematician, owes a lot to the presence of a third party outside of either discipline— the Office of Naval Research. Even back in 1992, ONR had discussed the possibility of creating a research initiative to bring the minds and skills of mathematicians and oceanographers together. The promise of such an initiative was a considerable incentive to us, not merely because it offered funding but also because it could mean recognition of the work.

Ironically, the initiative for such interdisciplinary work did not become a reality until late 1998, when ONR began funding projects under its new Departmental Research Initiative. But even during the years when the initiative remained merely an idea, it served as a catalyst. "It seemed for many years that we were working to chase something that always escaped us just as we reached it," Chris recalls. The potential for the ONR research initiative encouraged us would-be collaborators to begin the process of melding our research cultures. Yet because the initiative was not yet real, and ONR was not sending anyone checks, this critical getting-to-know-each-other phase unfolded casually. Talk of an initiative, in Chris's words, "gave us long-term hope without short-term pressure."

3. Wavelets: A Case Study of Interaction Between Mathematics and the Other Sciences
Ingrid Daubechies
Department of Mathematics
Princeton University

During the 1980s, an often-haphazard play starring mathematicians and scientists unfolded on many stages concurrently. Out of it emerged a versatile new mathematical tool—wavelets, which are being used by everyone from theoretical physicists and neuroscientists to electrical engineers and image-processing experts.

One of the early actors in this play is Alex Grossmann, a theoretical physicist at the Centre Nationale de la Recherche Scientifique, France, with a specialty in quantum mechanics. Through a common friend, he met Jean Morlet, a geophysicist at the oil company Elf Acquitaine, who was interested in finding better ways of extracting information about Earth from the echoes of seismic waves. At the time, Morlet was looking for alternatives to the workhorse mathematical technique of Fourier transformation for analyzing seismic data. The transformation was proving inadequate for combining the detail he desired at high frequencies with keeping track, simultaneously, of low-frequency behavior. Morlet intuited that there ought to be a different type of transform that would be able to reveal finer features in the data.

Grossmann recognized a connection between Morlet's proposed technique and his own work focusing on the coherent states of quantum systems. Together they hammered out the mathematics behind the first "wavelet transforms." Wavelet transforms are mathematical tools that enable researchers to decompose complex mathematical functions (or electronic signals, or operators) into simpler, easy-to-compute building blocks. There are other such tools out there, but not with the scope wavelets have. Wavelets are good building blocks to describe phenomena at different scales. Thus, sharply localized wavelets describe or characterize fine-scale detail of something like acoustic signals or neurophysiological data, while the more spread-out wavelet building blocks describe coarser features of the phenomena.

The connection Grossmann recognized between his work in quantum mechanics and Morlet's work in geophysics resided in the dualities inherent in the data of both fields. Quantum mechanics experts routinely confront the duality of a particle's position and momentum, which they only can determine precisely one at a time. To Grossmann, that was akin to the geophysical duality of time and frequency analysis of seismic and acoustic signals.

The link between these dualities and wavelet transforms subsequently emerged from the quantum mechanical concept of coherent states. Using these states as elementary building blocks, physicists can construct arbitrary state functions and operators even though each coherent state remains well localized around specified momentum and position values. A standard coherent state decomposition corresponds to the kind of windowed Fourier transform Morlet was familiar with. But a variant on this construction corresponded to the new transform Morlet had intuited ought to be possible, namely, a wavelet transform.

Word of this new way to process data began spreading, and more mathematicians began joining in. Yves Meyer, a renowned harmonic analyst, was one of them. He heard about Grossmann and Morlet's work while he was standing in line at a photocopy machine. He recognized their wavelet reconstruction formula as a rephrasing of an earlier construction by the mathematician Alberto Calderon, of the University of Chicago. That insight, in turn, moved Meyer to read an early wavelet paper jointly authored by Grossmann and me. And that input contributed to Meyer's own construction of a very beautiful and surprising wavelet basis. At the time, we would have bet, wrongly, that such a construction could not exist.

More scientists also began entering the fray. Stephane Mallat, a vision researcher at the University of Pennsylvania, heard about Yves Meyer's basis from an old friend of his, who had been a graduate student of Meyer's. Mallat realized that the hierarchical view prevalent in vision theory might feed back into wavelet theory and lead to a different understanding of a wavelet's construction. That hunch led to a collaboration with Meyer that yielded multiresolution analysis, a much better and more transparent tool for understanding most wavelet bases.

As a scientist who needed to understand and represent images in an algorithmic way, Mallat was motivated to relate the wavelet bases constructions with practical algorithms. It turned out later that those same algorithms already were in the hands of electrical engineers, who were using them in digital signal processing. Using these algorithms as a point of departure, I developed a series of different bases constructions in which the wavelets are defined through the algorithm itself. And that led to especially efficient wavelet-based computational techniques.

In the whole chain of events, which is only partially outlined here, several accidental encounters proved critical to the outcome—a mathematical framework for data analysis with widespread practical application whose development depended heavily on the input of scientists.

Many interesting questions arise in connection with this success story. What technical precursors provided entry for the idea that eventually became wavelet theory? What made these contacts between mathematicians and other scientists and engineers work so well? What sociological or other special factors, if any, played a role? Could this synergy have happened sooner? And what is happening now—are collaboration and cross-fertilization across disciplinary boundaries still taking place?

One of the reasons wavelet theory took off is because most wavelet transforms are associated with fast algorithms. That means that wavelet-based models of even real-life, large-size practical examples become computable. The situation is similar to the Fourier transform, which mathematicians and theoretical physicists have used for a very long time. The technique became a practical tool for scientists only after researchers made it more computable by developing the variant aptly known as fast Fourier transform. The same applies for wavelets. Although their mathematical content is useful regardless of the speed of algorithms for computing with them, most of their concrete applications would not exist if we didn't have a fast algorithm.

Because of its association with wavelet theory, the algorithm itself has been even more widely applied than ever. And that has cycled back into the mathematical development of wavelet theory. For example, wavelets led to a new mathematical understanding of the subband filtering algorithms first developed by electrical engineers. That new perspective, in turn, led to the construction of novel filtering algorithms designed to optimize certain mathematical properties of the associated wavelets. What's more, all of this attention to the original subband filtering algorithms ended up revealing many more applications for the algorithms than would have seemed natural from the filter point of view alone.

Wavelet theory has many more technical roots. Among them is approximation theory. As one builds up a function from its wavelet components, starting with the coarse-scale ones and then adding finer and finer scales, the construction is very similar to typical arguments in approximation theory, using, for example, successive approximations by splines. Aspects of vision theory and computer-aided geometric design also feed into the emergence of wavelets.

Simple as it may sound, however, the open-mindedness of human beings may have been the most important factor of all. Yves Meyer, for one, told me that his first reaction to the work of Grossmann and Morlet was, "So what! We [harmonic analysts] knew all this a long time ago!" That might have hampered the ascent of wavelet theory. But he looked again and saw that Grossman and Morlet had done something different and interesting. He built on that difference to eventually formulate his basis construction.

The open-mindedness of engineers has played an equally important role in the development of wavelet theory. Consider Martin Vetterli at Berkeley. Like Meyer, he overcame his first reaction, shared by many electrical engineers, that wavelets seemed but a rediscovery of their own subband filtering algorithms; when he looked at the wavelet papers more closely, he saw there was more.

Vetterli is now a human bridge between mathematicians and scientists and engineers. Skilled at explaining engineering issues to mathematicians, he also enjoys learning about mathematical ideas from mathematicians. What's more, engineers like Vetterli are the ones who tend to usher ideas like those from wavelet theory several steps closer to the "true" applications and who point out to mathematicians interesting mathematical developments by engineers.

Even when mathematical ideas turn out to be useful for specific scientific and engineering problems, they usually require further development to make them practical. In image compression, for example, thresholding (which in wavelet language translates into setting to zero all wavelet coefficients below a threshold) may be mathematically equivalent to more sophisticated approximation techniques. But in practice, engineers squeeze out much better results by using smarter procedures of bit allocation. Practical knowledge of this kind sometimes feeds all the way back to the mathematics of wavelet theory; in this particular case, the engineering approach to obtain a better coding led to interesting new approximation results. There is a kind of poetic justice to the way engineers' "smart tricks," as Vetterli calls them, turn out to be useful mathematical ideas in their own right that feed back into theoretical advances.

Interestingly, almost all the players in the early development and application of wavelet theory had one or more changes of field or subfield during their scientific lives. Exposure to different fields may be a key to becoming open to ideas from other fields. Or it may be that researchers who choose to develop expertise in more than one field also are the ones more inclined to listen to and explore ideas from other fields.

Technology, most notably easy access to computers, also was important in the development of wavelet theory in the 1980s. I probably would not have taken the path I did take had I been in my home lab in Brussels. At that time and place, access to computer facilities was not as straightforward as here. But I was visiting the Courant Institute at New York University, where I found a computer terminal on my desk.

The potential payoff in knowledge and technology was yet another incentive that accelerated the development of wavelet theory. That it might lead to new insight about vision inspired both Yves Mallat and me. When I wrote my first paper on the bases I constructed, I included a table of coefficients and a description of the algorithm. It laid out more clearly than usual, in a mathematics paper, how the construction could be used in practice. That opened up the technique even to those uninterested in slogging through the mathematical analysis of how I derived the particular numbers, or how I proved the mathematical properties of the corresponding wavelets.

Including a ready-to-use table of numbers rather than just a description of how these numbers could be derived should one choose to do so, was a feature borrowed from engineering articles. This was not a customary way of communicating in mathematical circles. So I am fortunate that *Communications in Pure and Applied Mathematics,* the math journal that published the paper, did not flinch at including such a lengthy table. It was a valuable tool. It is what inspired engineers and other scientists to try out the construction, even though the paper appeared in a journal that many of them would not routinely check.

When it comes down to it, those who communicate well to diverse audiences are the ones who have the most impact. The most influential papers by mathematicians in this area are written in a more transparent, more readable style than mathematics papers often are. In my own case, I was trained as a physicist even though I became a professional mathematician. That background has helped me communicate my own work in ways accessible to many scientists.

The culture of mathematicians often leads to work that appears opaque to those outside the culture. Many mathematicians write their journal papers in an extremely terse style. That succeeds in getting a maximal number of results onto limited journal pages, but it often keeps people outside the subfield from understanding the research. The lesson is simple: mathematicians who hope to collaborate with other scientists successfully must try to write their journal articles, and communicate in general, with their would-be collaborators in mind.

The synthesis of minds, ideas, and research that yielded wavelet theory in the 1980s probably could have happened sooner. Many conceptual precursors, including powerful mathematical ideas in harmonic analysis, had been developed in the last 40 years. But they remained largely confined to a small community. Similarly, the algorithms that underlie the fast wavelet transform were known almost only to mostly electrical engineers. Until the wavelet synthesis took place—almost by accident—it was not clear these ideas could be useful outside their own communities. The question never even came up, because the would-be suitors were completely unaware of one another. Had there been better mechanisms to promote interdisciplinary contacts, the wavelet synthesis might have happened at least a decade sooner.

In summary: the interdisciplinary drama that led to wavelet theory worked because it involved mathematicians who were open to other scientists and capable of explaining their work and ideas to nonmathematicians, and because of the involvement of engineers who were interested in interacting with mathematicians and had a similar open-mindedness. I think the most important lesson for mathematicians, however, is that we must value and cultivate this type of interdisciplinary contact just as we value intradisciplinary contacts between colleagues in different subfields of mathematics. Contacts like those have a knack for sparking new discoveries.

4. From Forest Dynamics to Interacting Particle Systems
Mercedes Pascual
Center for Marine Biotechnology
University of Maryland Biotechnology Institute
University of Maryland

The fundamental problems confronting ecologists today concern the rapid loss of biodiversity and its consequences for the functioning of ecosystems. Ecosystem function involves a variety of processes on which humans depend, such as nutrient cycling, responses to environmental perturbation, and mediation of global climate. Ecological modelers today are challenged to link biological structure to large-scale processes and, inversely, to link global environmental change to local effects on natural systems. Models that incorporate both a large number of variables as well as biological and physical processes at different, if not disparate, scales are called for.

The collaboration of Simon Levin and Steve Pacala (Princeton University) on forest ecosystems illustrates some of these challenges and underscores the broad expertise—from field ecology to theoretical ecology to mathematics—brought to bear on ecological modeling. The story of this collaboration also illustrates the key elements that enable such cross-disciplinary activities.

Pacala, a plant and theoretical ecologist, and Levin, a theoretical ecologist and mathematical biologist, share a common interest in forest ecosystems and in the scaling problems inherent in forest dynamics. In forests, the individual is the fundamental unit of ecological interaction and the natural scale at which to make measurements on demography, dispersal, etc. However, there is an enormous discrepancy between the temporal and spatial scales at which measurements are possible and those at which forest dynamics evolve for aggregate quantities such as total biomass or carbon and for processes such as the coexistence of species.

The initial discussions by Levin and Pacala on scaling of forest dynamics were motivated, in part, by a funding opportunity at DOE (although the work that followed was later cofunded by three other agencies). This work built on a stochastic and spatial model for forest ecosystems known as SORTIE, which followed the fate of individual trees. The initial version of SORTIE emerged from a group effort by Pacala and other plant ecologists (C.D. Canham, R.K. Kobe, and A.J. Silander) and serves today as the basis for modeling other forest ecosystems. The model was initially parameterized for Northeastern forests, raising challenging statistical questions on parameter estimation (Pacala et al., 1996).

Mathematics came into play in the subsequent efforts to simplify SORTIE. Simplification of forest simulators is needed to understand better what controls patterns, to achieve more robust parameterizations, and to embed forest dynamics in models that resolve physical space at larger scales (Levin et al., 1997). The complexity of the model is due to the tens of thousands of trees typically followed in a single simulation and to the large number of simulations needed to obtain the dynamics of aggregate statistical quantities (means, variances, and covariances) so that comparisons to field data are meaningful. Simulation studies of SORTIE with varying degrees of aggregation achieved significant simplification (Deutschman et al., 1997) and demonstrated the importance of biodiversity to the dynamics under enhanced carbon dioxide levels (Bolker et al., 1995).

Greater simplification was achieved, however, through analytical approaches. The result was a system of coupled partial-differential equations capable of accurately approximating the macroscopic dynamics of SORTIE.

The success of this derivation is apparent beyond SORTIE. A similar approach is being followed in the development of an ecosystem model to study terrestrial biological feedbacks on climate and climate change. The scaling problems are again daunting, from models based on physiological principles to prediction of relevant long-term community and ecosystem dynamics (Hurtt et al., 1998).

The linkages to mathematics of this work, however, go beyond the application of existing methods: they extend to developments in the area of probability theory known as interacting particle systems. This subfield originated in the 1970s, at which time it was motivated primarily by problems in physics. More recently, ecological systems and individual-based models like SORTIE have also been providing motivation. Examples can be found in the collaborative work of mathematicians Rick Durrett (Cornell University) and Claudia Neuhauser (University of Minnesota) with Levin and Pacala (e.g. Durrett and Levin, 1994; Neuhauser and Pacala, 1998). The mathematical work includes the study of connections between the behavior of spatial stochastic models and the behavior of differential equations.

The collaboration between Levin and Durrett began at a Cornell meeting, when Levin presented some results on ecological scaling that fit into a mathematical framework Durrett knew well. Weekly meetings followed. This follow-up was possible because both men were at Cornell at that time. Neuhauser, a graduate student of Durrett, first became interested in ecology when she took a seminar course taught by Levin. Later on, a Sloan Foundation fellowship enabled her to take 1 year off from her faculty position in the mathematics department of the University of Wisconsin to visit the department of Ecology and Evolutionary Biology at Princeton, where Levin and Pacala also had become faculty members.

The collaborative work described here was made possible by (1) scientists with strong quantitative training who played a pivotal role in the dialogue between fields and in the formulation of key questions, (2) mathematicians who were able to invest the time to explore connections to biology as a source of new problems, and (3) funding opportunities that stimulated the initial collaboration and provided support for graduate students and postdoctoral fellows. Pacala himself acknowledges that his NIH doctoral fellowship was instrumental in giving him the time and, accordingly, the freedom to learn more mathematics while pursuing graduate studies in ecology.

Bolker, B., S.W. Pacala, F.A. Bazzaz, C.D. Canham, and S.A. Levin. 1995. Species diversity and ecosystem response to carbon dioxide fertilization: Conclusions from a temperate forest model. Global Change Biology 1:373-381.

Deutschman, D., S.A. Levin, C. Devine, and L.A. Buttel. 1997. Scaling from trees to forests: Analysis of a complex simulation model. Science Online (available at <www.sciencemag.org/feature/data/deutschman/index.htm>).

Durrett, R., and S.A. Levin. 1994. The importance of being discrete (and spatial). Theoretical Population Biology 46(3):363-394.

Hurtt, G., P.R. Moorcroft, S.W. Pacala, and S.A. Levin. 1998. Terrestrial models and global change: Challenges for the future. Global Change Biology 4:581-590.

Levin, S.A., B. Grenfell, A. Hastings, and A.S. Perelson. 1997. Mathematical and computational challenges in population biology and ecosystems science. Science 275:334-342.

Neuhauser, C., and S.W. Pacala. Forthcoming. An explicitly spatial version of the Lotka-Volterra model with interspecific competition. The Annals of Applied Probability 9(4):1226-1259.

Pacala, S.W., C.D. Canham, J. Saponara, J.A. Silander, R.K. Kobe, and E. Ribbons. 1996. Forest models defined by field measurements: estimation, error analysis, and dynamics. Ecological Monographs 66(1):1-43.

5. Modeling the Dynamics of Infectious Diseases: Two Examples
Mercedes Pascual
Center for Marine Biotechnology
University of Maryland Biotechnology Institute
University of Maryland

Research at the interface of epidemiology and mathematics has a long and distinguished history, one whose beginnings are often credited to the work of Daniel Bernoulli on smallpox control in 1760. In the first part of this century, mathematical models for the spread of infectious diseases were pioneered by Ross, MacDonald, Kermack, McKendrick, and others. One fundamental contribution of that work was the threshold theory of epidemics (Kermack and McKendrick, 1927). This theory established that the introduction of a few infected individuals in a population will not give rise to an epidemic outbreak unless the number of susceptible individuals is above a certain critical value. Important public health problems motivated, and continue to motivate, research linking mathematics to epidemiology.

In the 1980s, interest in infectious diseases and epidemiological modeling increased exponentially. This increase was primarily due to the threat posed by acquired immunodeficiency syndrome (AIDS) but was also stimulated by the resurgence of penicillin-resistant gonorrhea and drug-resistant tuberculosis and malaria and by the emergence of new diseases such as Lyme disease, Legionnaire's disease, toxic shock syndrome, and hantavirus. Today many challenging mathematical and computational problems on modeling host-pathogen systems remain: not only are these systems highly nonlinear and stochastic, but they involve coevolutionary processes of hosts and pathogens, as well as spatial, genetic, and social heterogeneity (Levin et al., 1997).

Examples of effective interactions between mathematicians and epidemiologists can be found in the study of sexually transmitted diseases such as gonorrhea and AIDS. In the late 1970s and early 1980s, mathematicians Jim Yorke (University of Maryland) and Herbert Hethcote (University of Iowa) collaborated on the modeling of gonorrhea transmission and control. Yorke traces the origin of his involvement in this work to the proof of a theorem (with Ken Cooke) on differential delay equations. He became interested in the possibility of identifying a disease whose dynamics might be described by this type of equation; gonorrhea surfaced as a candidate disease during subsequent discussions with other researchers. But the key discussions leading to the research on the dynamics of this disease happened at a meeting of the Society for Industrial and Applied Mathematics. This meeting brought together Jim Yorke, Herbert Hethcote, and Rafe Henderson from the Centers for Disease Control and Prevention (CDC).

The research by Yorke and Hethcote that followed was possible because of the cooperation and encouragement of Rafe Henderson and Paul Wiesner, Directors of the Sexually Transmitted Diseases Division at CDC. The work was supported for 5 years by grants from the CDC and the National Institutes of Health. The resulting models were described as "extremely useful in formulating approaches to gonorrhea control at the national level" (Wiesner and Cates, 1984). Furthermore, the particular model developed and analyzed by Yorke and graduate student Lajmanovich and Yorke (1976) stimulated mathematical work by Morris Hirsch, Hal Smith, and others on monotone dynamical systems that are competitive or cooperative (e.g.,

Hirsch, 1984). Morris Hirsch acknowledges the "powerful influence" of the epidemic models on his past and current research on dynamical systems (e.g., Benaïm and Hirsch, 1997).

A more recent example at the interface of mathematics and epidemiology is the collaborative work of epidemiologist Jim Koopman, mathematician Carl Simon, and physiologist/mathematical modeler John Jacquez on the transmission of the human immunodeficiency virus (HIV), which leads to AIDS. When asked what made this collaboration effective, Jacquez emphasized the willingness of each of the collaborators to learn about the others' fields. Based on models they developed for sexual partnership formation, and on their work to fit data on HIV and AIDS within a major risk group, the Michigan group concluded that most of the transmission of HIV occurs in the very early, preantibody period of infection (Jacquez et al., 1994; Koopman et al., 1997). The 1994 paper addressing this theory was awarded the Howard Temin Prize in Epidemiology for the best paper in epidemiology in the *Journal of AIDS*.

It is interesting to consider that this interaction started in 1986 and grew over a considerable period of time. Thus, even at an interface of mathematics and science that has a long history, collaborations can require fairly long times—relative to the average funding cycle—to become established and flourish.

Benaïm, M., and M. Hirsch. 1997. Differential and stochastic epidemic models. The Fields Institute Communications. Vol. 21: Differential Equations with Applications to Biology. Proceedings of the International Conference on Differential Equations with Applications in Biology held in Halifax, Nova Scotia, Canada, July 25-29.

Hethcote, H.W., and J.A. Yorke. 1984. Gonorrhea Transmission Dynamics and Control. Lecture Notes in Biomathematics. New York: Springer-Verlag.

Hirsch, M. 1984. Stability and convergence in strongly monotone dynamical systems. Journal reine angew. Math. 383:1-53.

Jacquez, J.A., J.S. Koopman, C.P. Simon, and I.M. Longini. 1994. Role of the primary infection in epidemics of HIV infection in gay cohorts. JAIDS 7:1169-1184.

Kermack, W.O., and A.G. McKendrick. 1927. A contribution to the mathematical theory of epidemics. Proc. Roy. Soc. Lond. A 115:700-721.

Koopman, J.S., J.A. Jacquez, C.P. Simon, B. Foxman, S. Pollock, D. Barth-Jones, A. Adams, G. Welch, and K. Lange. 1997. The role of primary HIV infection in the spread of HIV through populations. JAIDS and HR 14:249-258.

Lajmanovich, A., and J.A. Yorke. 1976. A deterministic model for gonorrhea in a nonhomogeneous population. Mathematical Biosciences 28(1976):221-236.

Levin, S.A., B. Grenfell, A. Hastings, and A.S. Perelson. 1997. Mathematical and computational challenges in population biology and ecosystems science. Science 275:334-342.

Wiesner, P., and W. Cates. 1984. Foreword to Gonorrhea Transmission Dynamics and Control: Lecture Notes in Biomathematics. New York: Springer-Verlag.

6. Topology and Dynamics of Mutant Bacteria, and Applications to Materials Science
Michael Tabor
Department of Mathematics
University of Arizona

Neil Mendelson is a molecular biologist at the University of Arizona who became interested in the properties of a particular mutant strain of the bacterium *Bacillus subtilis*. This strain has the unusual property that the growing cells lack the ability to separate, with the consequence that they form long filamentary structures exhibiting remarkable dynamical behavior. Subtle features of the bacterial cell wall structure impose a specific handedness and twist on the growing system. After sufficient replication-induced growth, and as a result of dramatic flailing motions resulting from the built-in twist, the end points of the bacterial strand meet, leading to the formation of a closed loop, which then winds itself up into a double-stranded helical structure that maintains its original handedness. Eventually, owing to continuing cell replication, the growing double-stranded filament repeats the same process and becomes a quadruply twisted fiber; ultimately, after many repetitions of this remarkable dynamics, it forms a macrofiber, a fiber of macroscopic dimensions.[1]

Mendelson's pursuit of this nontraditional research topic led to a decline in scores on his grant applications, loss of funding and, for a time, significant damage to his publication record. However, chance interactions with engineers, mathematicians, and physicists, all with an interest in elasticity theory, eventually led to some remarkably successful interactions and collaborations across traditional disciplinary boundaries. For example, some years ago, he started a collaboration with an engineer, John Thwaites, from the University of Cambridge, whose expertise in fiber science enabled them to undertake detailed studies of the basic mechanical properties of the bacterial filaments (Thwaites and Mendelson, 1991).[2]

Mendelson's collaboration with mathematicians started about 6 years ago with the arrival at Arizona of Michael Tabor, who had been recruited from Columbia University to become head of the University's renowned Applied Mathematics Program. Tabor was keen to develop interactions between biologists and mathematicians and had a personal interest in elasticity theory. A chance meeting with Mendelson soon after his arrival marked the beginning of a particularly fruitful interaction in which the results of the experimental biology stimulated the development of a whole new set of mathematical results (Goriely and Tabor, 1999). These have included the formulation of (the first) dynamic models of flexible elastic filaments[3]—models that have demonstrated how "writhing"[4] stems from subtle mathematical instabilities within the system. These results are not restricted to the biological problem that stimulated them, and the new mathematical formalism has since been applied to describe the behavior of solar flux tubes, which emerge from the Sun's interior as narrow strands of intense magnetism and appear as sunspots (Longcope and Klapper, 1997). Another recent application has been to the mechanics of climbing plants (Goriely and Tabor, 1998).

---

[1] Mendelson published his first paper on this topic in the Proceedings of the National Academy in 1976. A more recent review can be found in Mendelson (1990).
[2] This is one of a number of papers published by Thwaites and Mendelson.
[3] Devised by Tabor in collaboration with his postdoctoral associates Alain Goriely (now on the faculty at the University of Arizona) and Isaac Kapper (now on the faculty at Montana State University).
[4] Writhing is a measure of the nonplanar deformation of space curve.

More recently, the arrival of a new member of the Physics Department, Ray Goldstein, has led to further interactions, and Mendelson's system has inspired Goldstein, who has a strong interest in biophysical research problems, to devote a significant proportion of his research program to studying the physical properties of the bacterial filaments. As with the work of Tabor and his collaborators, Goldstein's research has fed back ideas to Mendelson and stimulated new ways of thinking about the biological process itself.

According to Mendelson, one of the things that enabled the extensive interdisciplinary interactions to occur was the ability to pose the problem in the right way. In this case, the bacterial system was videotaped so that the mathematicians could see the dynamics directly and develop a geometric intuition about the filament dynamics. Equally important was the mutual respect and curiosity that the participants had for each others' disciplinary perspectives and their willingness to risk devoting a significant amount of time to understanding each others' ideas and methodologies.

Mendelson's system is not just a biological curiosity: it is also of interest to materials scientists because of the macrofiber's special spatial structure and its capacity to bind with various minerals, either by forming very strong, lightweight fibers or by serving as templates for porous silicates (Mendelson, 1996). Mendelson has observed that the bacterial filaments bind to mineral salts in solution, a process that stiffens the threads, transforming them from flexible macrofibers to crystalline fiber networks. These bacterial-mineral networks, called "bionites," can be color-selected according to the identity of the mineral component; they are biodegradable and stronger than steel (on a per-mass basis). Further, bionites have been shown capable of serving as templates for medical materials. Materials scientists have since demonstrated the ability to synthesize siliceous bionites and then destroy the bacterial components without significant shrinkage of the material. Stephen Mann, in particular, has demonstrated the possibility of producing silicon implants having sufficient porosity to allow for the attachment of natural tissue (Davis et al.,1997).

In the last few years Mendelson's work has become better recognized. He was invited to speak about his research at the 1996 conference of the American Society for Microbiology (ASM)[5], and *Science* (1997) has featured his work and multidisciplinary collaborations. Nonetheless, despite the increased scientific recognition, research funding is still difficult to obtain.

Davis, S.A., S.L. Burkett, N.H. Mendelson, and S. Mann. 1997. Bacterial templating of ordered macro-structures in silica and silica-surfactant mesophases. Nature 385:420-423.

Goriely, A., and M. Tabor. 1997. Nonlinear dynamics of filaments I: Dynamic instabilities. Physica D. 105:20-44.

Goriely, A., and M. Tabor. 1997. Nonlinear dynamics of filaments II: Nonlinear analysis. Physica D. 105:45-61.

---

[5]Having grown in membership from 59 scientists (1899) to over 40,000, the ASM is the oldest and largest life science membership organization in the world.

Goriely, A., and M. Tabor. 1997. Nonlinear dynamics of filaments III: Instabilities of helical rods. Proc. Roy. Soc. 453:2583-2601.

Goriely, A., and M. Tabor. 1998. Spontaneous helix hand reversal and tendril perversion in climbing plants. Phys. Rev. Lett. 80:1564-1567.

Goriely, A., and M. Tabor. 1999. Nonlinear dynamics of filaments IV: Spontaneous looping of twisted elastic rods. Proc. Roy. Soc. 455:3183-3202.

Longcope, D., and I. Klapper. 1997. Dynamics of a thin twisted flux tube. Astrophysics J. 488:443-453.

Mendelson, N. 1990. Bacterial macrofibers: the morphogenesis of complex multicellular bacterial forms. Sci. Progress 74:425-441.

Mendelson, N. 1996. Bacterial fibers and their mineralized products bionites. pp. 279-313 in Biomimetic Materials Chemistry, S. Mann, ed. New York: VCH Publishers.

Physics, biology meet in self-assembling bacterial fibers. 1997. Science 276:1499-1500.

Thwaites, J., and N. Mendelson. 1991. Mechanical behavior of bacterial cell walls. Adv. Macrobiol. Physiol. 32:174-222.

7. Examples from Molecular Biology
Nicholas Cozzarelli
Department of Molecular and Cell Biology
University of California at Berkeley

My own attempts to apply mathematics to a problem in biology were initially frustrating but ultimately successful. They provide some lessons that are useful in other cases. They also ultimately led to the creation of a national program to facilitate the interactions between mathematicians and molecular biologists.

Topology of DNA Knots and Links

In 1979, while at the University of Chicago, I was working on the mechanism and function of topoisomerases, the essential enzymes that control the topology of DNA in all cells. The winding of the two strands of the double helix about each other is described by the topological property known as the linking number. We determined that topoisomerases changed the linking number of DNA either by 1 (type 1) or 2 (type 2) in a single step. The mechanism of type 1 and type 2 enzymes must therefore differ in a pronounced way. As a biochemist, I had only the most tenuous grasp of the meaning of linking number and therefore of what might be the implications of our discovery. I tried to read papers on topology but could not get beyond the first paragraph, because the concepts, background, and language were totally foreign.

A further impetus for learning DNA topology came in a letter from a topologist, William Pohl, who said that I had quoted the topology literature inaccurately in our linking number change paper. I pride myself on referencing fairly, but I could not justify my summary of the literature. I went to the University of Chicago Mathematics Department and tried to explain the issues to a topologist. He was very kind and went with me to the library to look up the mathematical papers on linking number but there was no useful outcome. I could not explain to him in precise terms what I needed to know and he could not identify the critical issues as he knew essentially no biology—his last exposure had been in high school. (I know now that neither the references I had used nor the ones that Pohl had suggested were totally correct, but I only found this out 15 years later, when I knew how to frame the questions better.)

Several years later when I was at the University of California at Berkeley, my group made another discovery that demanded even more collaboration with a topologist. We found that coating DNA knots and links with a particular protein made the path of the DNA traceable by electron microscopy. For the first time one could determine the topology of DNA knots and links. We tried to determine which kind of knots were produced by topoisomerases and recombinases (enzymes that reshuffle the DNA sequence) in the test tube and in cells. Our method for determining knot type was crude. We took a picture of the knot or link by electron microscopy and folded a plastic tape into the same path. We then twisted the tape until it seemed to have the simplest form. We drew that representation and compared the drawings.

One day, after a student of mine and I had classified a large number of complex knots and links in this way, I inadvertently knocked the plastic tape onto the floor. It adopted a new and even simpler path, which showed that this knot was the same as a knot we had seen before; it wasn't a new knot. Thereafter, we routinely threw the plastic tape on the floor so that we could approximate a "lowest energy level." This procedure worked fairly well for simple knots and links but was useless for more complex ones. Just because you are unable to twist a plastic ribbon representation of one knot to be like another knot does not mean the knots aren't still identical.

Once again I went to a mathematics department—this time at Berkeley—and sought out a topologist. The result was the same as at Chicago. I could not formulate the questions to him in a way that he understood.

I decided to write to three mathematicians who had worked with DNA and to ask them for help. I hoped that their knowledge of biology would bridge the gap that existed between me and the mathematicians. James White answered my letter and said that topologists could indeed solve my problems. He introduced me to the world of topological invariants and to topologists—including Ken Millet, Lou Kauffman, Vaughan Jones, and De Witt Sumners—who worked with them. This started some productive collaborations in which we explained to the biologists how to use invariants and used them ourselves to study DNA structure and transformations. It was a two-way street. The mathematicians were fascinated by the rich variety and essentiality of DNA topology and were often surprised at the forms they adopted when viewed by electron microscopy.

In one paper I went too far in the use of topology, but it brought me to a new way of looking at DNA. We had found something very interesting about the effects of catenating two rings on the linking number of each ring and had interpreted the results in terms of surface topology. The data were correct but the interpretation was not useful. We received a dozen letters from around the world pointing this out and suggesting various alternative interpretations. I had by that time become director of the Program in Mathematics and Molecular Biology (PMMB), and I invited all of our critics to a workshop at a meeting of PMMB in Santa Fe, New Mexico.

At that workshop, most of us concluded that surface topology was great for describing some biological processes but that another formulation was simpler and more useful in describing our results with supercoiling. One of the most vociferous critics of the use of surface topology to describe supercoiling was a Russian, Alex Vologodskii, who was an expert in the use of computer methods to simulate DNA. At the Santa Fe meetings, Alex and I realized that our approaches were nicely complementary. I was an experimentalist who was uncovering the roles and uses of DNA topology and he was a theoretician who could, via simulation, test out the underlying bases for our results. We started a collaboration that continues to this day. A number of important results have emerged that would not have been possible had each of us been working alone. Let me give just one example.

A vexing but important problem in biology is the conformations of supercoiled DNA. The DNA in all organisms is supercoiled, and proper supercoiling is vital to function. Indeed, drugs that block topoisomerases, which introduce and remove supercoils, are among the most widely used antibacterial and anticancer agents. The development of these new drugs was a direct consequence of the basic studies on topoisomerase mechanism and structure. DNA must be relatively large to be supercoiled, and this precluded the use of high-resolution structural methods, such as X-ray crystallography. Instead, we found that a combined approach—biophysical measurements and simulation—was very powerful. Because of the excellent agreement between theory and experiments, we could calculate what we could not measure. As a result, we could specify all the important conformational properties of DNA and how they varied with DNA length, supercoiling density, and the concentration of mono-, di-, and trivalent neutralizing ions. The problem that had seemed intractable was solved.

Program in Mathematics and Molecular Biology

Thus far this narrative has focused on my own work. In 1986, DeLill Nasser, an administrator from NSF who was involved in science and technology centers (STCs), called to encourage me to think about a center for mathematics and biology. I asked Sylvia Spengler at Berkeley if she might be interested in leading such a facility with me. She loved the idea and we set out to find like-minded mathematicians and biologists. Jim White readily joined, as did Jim Wang, Lou Kauffman, Vaughan Jones, and De Witt Sumners. We went to DeLill and said we were ready to submit our application to NSF, but she wisely counseled us to broaden out beyond DNA topology and geometry. In the next phase we brought Eric Lander, Mike Waterman, Gene Lawler, and Mike Levitt into the group. These were the founding members of PMMB, and NSF funded our program.

Our main job was to bridge the barriers between the two disciplines. To do so we held a workshop at Santa Fe like the one described above, where the audience was an equal mix of mathematicians and biologists. We had didactic lectures about mathematics for the biologists and about biology for the mathematicians. There were also cutting-edge lectures where the only rule was to try to be clear to both disciplines. We also left a lot of time for informal discussions. It worked remarkably well. In a survey we found that fully 35 percent of the participants in this one conference had formed an interdisciplinary collaboration.

But this was just one aspect of PMMB. Another was our matchmaker function, getting the right mathematicians and biologists together. Mathematicians and biologists, even those on the same campus, usually do not know each other, but we in PMMB had a lot of experience in interdisciplinary work and we had contacts in both communities. We invited biologists who were looking for a mathematician collaborator to ask us for a potential match. Likewise, mathematicians who were thinking of making the plunge into biology were invited to talk to us. We then tried to make appropriate matches or bring them to Santa Fe to find their own mates. I have no doubt that several years of work could have been saved if I had from the beginning been able to find the appropriate mathematical collaborators, and I did not want others to repeat that experience.

The whole arrangement would have little future unless we brought in young people. We did so through our fellowship programs. There were PMMB fellows in the members' laboratories and offices and a larger national fellowship program to extend our reach. The latter has been greatly expanded by funds from Burroughs Wellcome Foundation. Our requirement is just that the work we support is high quality and interdisciplinary. We also bring the young people to Santa Fe and to our retreats, so that they become part of a community.[1]

---

[1] The current members of PMMB are Bonnie Berger (Massachusetts Institute of Technology), Pat Brown (Stanford University), Carlos Bustamante (University of California, Berkeley), Nicholas Cozzarelli (University of California, Berkeley), David Eisenberg (University of California, Los Angeles), Vaughan Jones (University of California, Berkeley), Lou Kauffman (University of Chicago), Eric Lander (Massachusetts Institute of Technology), Mike Levitt (Stanford University), Wilma Olson (Rutgers University), Terry Speed (University of California, Berkeley), Sylvia Spengler (Lawrence Berkeley National Laboratory), De Witt Sumners (Florida State University), Elizabeth

In addition to the interdisciplinary research conference at Santa Fe, PMMB sponsors smaller workshops, short courses, and symposia at universities and professional society meetings. These activities result in text materials and collaborations between PMMB members. The number of highly qualified fellowship applicants has tripled since the PMMB's inception, indicating that the combined field of mathematics and biology is growing rapidly.

Thompson (University of Washington), Ignacio Tinoco (University of California, Berkeley), Jim Wang (Harvard University), Mike Waterman (University of Southern California), and Jim White (University of California, Los Angeles).

8. Challenges, Barriers, and Rewards in the Transition from Computer Scientist to Computational Biologist

Dannie Durand
Department of Molecular Biology
Princeton University

Metal, glass, and many other materials get stronger when they're heated up and allowed to cool slowly. The process, called annealing, works by enabling the materials' molecular structures to assume more stable arrangements than faster cooling rates would allow. In effect, annealing lets the materials "compute" an internal structure corresponding to an energy minimum. Simulated annealing is a computational technique corresponding to the physical one, which means it is a marriage of mathematics and physics. It has become a widely used heuristic method to solve combinatorial optimization problems that are too large to be solved exactly. The method has been particularly useful for optimizing electronic VLSI (very large scale integrated) circuit layout, which includes challenges like distributing individual components on a chip so that the total length of wire connecting them is minimized.

My own doctoral thesis in 1990 focused on the challenge of making simulated annealing calculations faster using parallel computers. After completing my doctorate, I worked for several years on a variety of scheduling problems in parallel computation in an industrial research laboratory before making the transition to another area—computational biology—where mathematics and science come together in a rich intersection.

I had been interested in computation and biology for some time, but my participation in the 1994-1995 Special Year on Mathematical Biology (hosted by the Center for Discrete Mathematics and Theoretical Computer Science, or DIMACS, at Rutgers University) helped me to begin research in this area. This special year focused on computational and mathematical problems that arise from the wealth of protein and DNA sequence data that has become available since the 1970s. Taken together, the tutorial, workshops, and seminars associated with the special year provided a comprehensive background in almost all areas of computational molecular biology. I also met many of the leading researchers in the field, became familiar with their published work, and learned about the field's important open problems.

My introduction to the field's players led to a collaboration with Lee Silver, a molecular biologist from Princeton. This was another event that was crucial for my transition from simulated annealing studies to research in computational biology. Silver and I met as part of the DIMACS Special Year, where a number of Princeton biologists presented computational problems that had arisen in their research. I became interested in his work because the biological system he described—a gene that promotes itself at the expense of the organism that carries it and the other genes in that organism—was bizarre and compelling. We began to meet weekly to discuss the problem.

Our initial collaboration addressed issues in the evolution and population dynamics of the *t* haplotype, a mutant chromosomal region in mice that violates Mendel's laws and causes sterility in males. Our first results on *t* haplotype (or just *t*) population dynamics were based on Monte Carlo simulations. They overturned the long-standing hypothesis that the low-level persistence of *t* is due to a balance between loss of *t* due to stochastic effects in small populations and reintroduction of *t* through migration. This work appeared in the journal *Genetics* in 1997.

As I learned more about this biological system, I saw that there were several interesting mathematical aspects as well. Two papers on the evolution of the *t* haplotype—one based on Monte Carlo simulation, the other on Markov chain analyses—are currently in preparation as a result. The main contribution of these papers is in the field of theoretical biology. The work also suggests new experimental hypotheses that may now be tested through laboratory and field work.

I am now working on a second project with Silver and one of his graduate students, Ilya Ruvinsky, that studies rearrangements in the mouse genome. Ruvinsky initially suggested this second project, which has a substantial bioinformatics component. Neither one of us has the expertise to pursue the project alone, and our collaboration would probably not have come to pass had I not been regularly attending Silver's lab meetings. If successful, this research will add to biology by shedding light on the evolution of mammalian genomes. It could contribute to computer science in the areas of Internet information retrieval, data mining, and data visualization.

Silver and I have an effective collaboration because each of us is willing to go to considerable lengths to understand the other's field and culture. He has enough computational expertise to understand what I have to contribute. He can also identify computational problems that arise in his laboratory and has a realistic sense of what it would take to solve them. I understand enough of the molecular biology to see just how computational methods can help solve problems and also understand the limits of those methods.

My switch to research in computational biology has been the most important decision in my professional career. My research has become much more rewarding and I have become far more productive. But there also are barriers to research in areas like computational biology that intimately combine mathematics and science. One barrier centers on how research in computational biology should be evaluated and rewarded. Can a computer scientist in a computer science department receive tenure if she publishes a significant part of her research in biology journals? Her colleagues may not value such work. Even if they do, they probably do not have the expertise to distinguish top quality work in computational biology from lesser efforts. Alternatively, will a biology department be willing to hire a professor with a PhD degree in computer science and little or no formal training in biology?

Another challenge to interdisciplinary programs is the physical separation of collaborators. On the one hand, researchers in an interdisciplinary program need to interact. For this to work, they need to be sitting in the same building and sharing the coffee machine, secretarial staff, etc. On the other hand, researchers also need to spend time in their own departments to stay up to date on new developments in their own fields. I, for one, select the problems to work on based on their biological importance. To solve these problems, I frequently need to draw on expertise from subareas in computer science other than my own. I can be most effective, therefore, if I spend a substantial part of my time sitting in a computer science department whose faculty members are interested in a fairly broad spectrum of areas.

It takes more than individuals with an interest in merging math and science before anyone actually does the interdisciplinary research that interest suggests. It requires a commitment to interdisciplinary research all the way from the national level, where funding agencies decide where to put their money, down to the university level and the level of individual departments. It requires infrastructure to support activities in both the home departments of collaborators and the interdisciplinary research center (or other venue) supporting the collaborative work. And it requires a mechanism by which to reward research activities that are specifically interdisciplinary.

9. Crossbreeding Mathematics and Theoretical Chemistry: Two Tales with Different Endings

R. Stephen Berry
Department of Chemistry
University of Chicago

Spectra of Nonrigid Molecules

The spectra of molecules—the pattern of wavelengths the molecules absorb or emit—reveal to us the structures of molecules and how they rotate, vibrate, and change their shapes. Interpreting molecular spectra is not always straightforward, and sophisticated mathematical tools frequently come into play when chemists extract the hidden messages of such spectra. Nonrigid molecules, whose atoms may exchange sites on time scales approaching the fast time scales of atomic vibrations in molecules, have spectra that pose some of the greatest challenges to interpretation. The effective symmetry of such a molecule appears higher than just that of the rigid ball-and-stick object we would normally use to represent it. When the atoms rapidly exchange places in the molecule, the result looks just like the original molecule (so long as identical atoms are unlabeled and hence indistinguishable) but is rotated in space from the original orientation.

Even as the atoms are switching their positions, the entire molecular framework also rotates at a comparable rate. The problem is much like that of how a falling cat rights itself as it falls. How do those atomic permutations interact with that normal, "rigid-body" rotation of the molecule and how does that influence the spectrum? And how can chemists infer from the pattern of the spectrum how the atoms move when they exchange places? One natural path to the solution is in the mathematics of group theory.

The part of group theory most relevant here is the theory of representations of continuous groups (Lie groups). More particularly, subgroups of the rotation group in three dimensions provide a natural way to describe the motions of nonrigid molecules. My chemist colleagues and I set off in that direction and soon bumped into some mathematical difficulties—specifically, the need to handle infinite, discrete groups (discrete subgroups of Lie groups).

We needed help. So I, as the leader of the research group, a chemistry professor, approached a member of the resident math faculty and explained our difficulty. The next day, the mathematician came back to say that he had found the problem interesting. More than that, he said it had led him to a conjecture in number theory. He then described the problem to a local number theorist, who proved the conjecture the next evening and wrote out his proof. Copies were sent to the other mathematician and, kindly enough, to the group of chemists whose search for help sparked this bit of innovation in number theory. The sad and ironic ending—or at least pause—for the chemists in this story is that the problem in group representation theory whose solution would enable them to interpret spectra went unsolved and remains so.

This comingling of scientists and mathematicians started out with promise, but it withered. Even a cursory postmortem reveals some of the pitfalls and challenges that members of these two professions confront when they aim to collaborate. One key factor is probably that the mathematicians and the theoretical chemists never came back together after the number theorist found the solution to the conjecture. There was neither a sufficiently strong collaborative sensibility on either side nor enough motivation for the mathematicians to return to the original problem. Although the mathematicians found the problem in number theory to be interesting enough at first, they soon deemed it more or less a dead end. Had the problem seemed to the mathematicians to have better long-term prospects, they would probably have wanted to pursue the connection the theoretical chemists had brought up.

One obvious lesson for scientists looking to work with mathematicians is that their would-be collaborators are more likely to become committed stakeholders in a project if they have reason to believe they will have something to gain in the long term. That might have happened eventually in this case, but the interaction disintegrated too soon to find out. There's a lesson in that too. Unless someone works at nurturing the nascent linkage between scientists and mathematicians, it is likely to fall apart.

Finite-Time Thermodynamics

Now for some better news. An undergraduate who had received a bachelor's degree with a major in mathematics went to graduate school in a chemistry department, intending to study theoretical chemistry. He fortuitously joined a group just at a time when a new problem had arisen that was tailor-made for collaborative research with mathematicians. The problem was in the area of thermodynamics. No one in the group had worked in this area before, but they took on the problem anyway. The question they asked themselves was this: Could they extend methods of thermodynamics to systems constrained to operate in finite times or at nonzero rates? That it was a math-heavy challenge was clear to everyone in the group. Fortunately, the requisite mathematics, especially differential geometry, was familiar and friendly territory to the new grad student. With his background in the intellectual mix, the group worked productively to show how a sophisticated use of mathematics could lead to new insights in theoretical chemistry, and in nonequilibrium thermodynamics in particular. One thing the group was able to do was articulate and then prove an existence theorem (with necessary and sufficient conditions) for quantities analogous to traditional thermodynamic potentials but for systems and processes with time constraints. In doing this, they solved the first major problem they set out to solve. The group also developed a procedure for generating these potential-like quantities via Legendre-Cartan transformations, a modern extension of the Legendre transformation widely used in natural sciences. In addition, they applied more mathematics—in this case, optimal control theory—to find the pathways that yield or best approach the optimum performances of thermodynamic processes, the performances given by the changes of the limiting potentials. The research turned out to be remarkably successful. The scientific knowledge of the professor and postdoctoral associates combined well with the graduate student's ability to learn and master new mathematics quickly. What had been merely a problem in thermodynamics grew into an entire subfield. It has even led to some new mathematics, an example of which concerns the properties of surfaces of constant but nonzero curvature. The then-student has become one of those human bridges whose hybrid knowledge and interests spawn interdisciplinary research. He became a faculty member of a Department of Mathematics, where he currently studies problems on the border between mathematics and the natural sciences.

The collective moral of these two vignettes is simple. Mixing science and math can be a formula for discovery in both mathematics and science. Achieving a successful collaboration, however, requires not only a serendipitous coincidence of ideas, interests, and recognition of relevance but also plenty of perseverance.

10. Martingale Theory
John Lehoczky
Department of Statistics
Carnegie Mellon University

The mathematical term "martingale" has been traced to the French village of Martigues, whose inhabitants were said to have a passion for gambling. The mathematician Jean Ville first used the term to refer to a particular type of stochastic process, which he applied initially to model a sequence of fair games. The martingale theory that developed out of this initial work did not remain exclusively within the realm of mathematics. It also would provide insights in many scientific areas. Some of those, in turn, would lead to new developments in mathematics.

In the language of mathematics, a martingale is a sequence of pairs $\{(X_n, \mathcal{H}_n)\}$, in which the random variable X_n represents the cumulative winnings of a gambler after n trials and $\mathcal{H}_n$ (a sigma field) represents the history of the gambler's fortunes up through the first n trials. The defining martingale property is given by the following equation:

$$E(X_{n+1} \mid \mathcal{H}_n) = X_n$$

from which one can derive

$$E(X_m \mid \mathcal{H}_n) = X_n$$

for all $m \geq n$. In words, the equations say that the cumulative winnings, on average, will remain unchanged over the sequence of future gambling trials.

Ville defined the martingale concept to model gambling. It carried little importance beyond that until the 1940s, when J.L. Doob and Paul Levy separately investigated the structure of martingale processes. They proved a variety of mathematical results, including limit theorems governing the processes' asymptotic behavior. More importantly, they generalized the concept of martingales to submartingales and supermartingales, which they linked to an entirely different branch of mathematics—the subharmonic and superharmonic functions arising in potential theory.

Doob worked out much of the interrelationship between martingale theory and potential theory, and the role that Brownian motion plays in the solution of certain partial differential equations. He summarized this research and provided extensive historical commentary in his 800-page book *Classical Potential Theory and Its Probabilistic Counterpart*, published in 1983.

That accelerated the diffusion of martingale theory into a wider range of mathematical areas and from there into scientific and technical areas as well. Consider the following vignette.

During World War II, a group of statisticians, including Abraham Wald, were working at Columbia University on applications of the mathematical sciences to military problems. One of those problems concerned the development of efficient sampling plans for inspecting materiel. Already in place were well-defined military standards for sampling materiel before shipping it to the battlefield. These standards specified such details as the number of items to be inspected and the actions to be taken for different test outcomes. For certain types of materiel like ammunition, the testing procedure was destructive so there was plenty of incentive to minimize the sample size. This is where martingale theory enters the picture.

Wald and his collaborators approached this sampling size issue by considering sampling plans with a variable number of units being tested. The number of units would depend on the earlier tests. For example, a sufficient number of early successes could allow one to infer that the shipment was of sufficient quality to accept the entire lot without wasting any further parts to testing.

Wald developed a particular testing procedure, now known as the Wald Sequential Probability Ratio Test (SPRT). In determining the performance of his procedure, he used martingale methodology, and his methods, in turn, made further contributions to martingale theory. The formalization of that concept eventually would influence potential theory through the notion of Brownian motion hitting the boundary of analytic sets. More importantly, Wald and Wolfowitz were able to prove the ability of the SPRT procedure to minimize the expected sample size for destructive inspections. This research would become the foundation of a new field of statistics, now known as sequential analysis.

Sequential analysis, in turn, evolved extensively and has proven valuable in many scientific and technical domains, including clinical trials in medicine. These techniques also have become important to designers of industrial experiments. Determining settings of an industrial process's different control parameters to optimize the output or yield is one important example.

Interestingly, the study of the performance of the SPRT also stimulated the development of another branch of mathematics—dynamic programming—which now falls within the larger framework of operations research. A seminal paper by Arrow, Blackwell, and Girshick, which offered an alternative proof of the SPRT's optimal performance, often is credited with starting the field of dynamic programming. Martingale theory also plays an important role in stochastic control theory.

Martingales have continued to serve scientists in a variety of ways that have circled back into the mathematical sciences. One example of this emerged in the 1980s from the work of Harrison, Kreps, and Pliska, who put martingale theory to work in financial mathematics. They followed research by others (among them Black, Scholes, Merton, and Samuelson) on the pricing of contingent claims.

The newer work developed the theory of continuous trading and revealed its connection to the absence of arbitrage possibilities. And this implied that discounted asset prices must behave like martingales. Then, by applying martingale theory developed many years earlier, it became possible to determine the arbitrage-free price of a contingent claim and how to reconstruct a portfolio to hedge it. The celebrated Black-Scholes formula is a special case of this methodology. Indeed, contingent claims prices often can be expressed as an expected value with respect to a "martingale measure"—a probability measure under which the discounted asset price process forms a martingale.

This particular marriage of mathematics and the financial world has had an important role in the creation of financial markets with a notional value today of trillions of dollars. More importantly, the connections between martingale theory and contingent claim pricing have raised a whole new set of problems in martingale theory for mathematicians to ponder.

Martingale theory continues to make marks in both mathematical and scientific arenas. The theory has been applied to counting process theory. That, in turn, has found use in the part of operations research known as queuing theory, as well as in survival analysis in biostatistics. What's more, Radon-Nikodym derivatives and likelihood ratios now can be examined in terms of martingales, which means martingale theory has earned an important place in mathematical statistics.

The interaction between mathematical science and the other sciences catalyzed by martingale theory clearly has borne much fruit for all parties involved.

B

Workshop Agenda and Presentations

AGENDA

March 25, 1998

8:00 a.m.	Chairman's welcome and introductions—Thomas Budinger
8:30 a.m.	Introductory remarks—Phillip Griffiths
8:45 a.m.	The Elucidation and Quantification of Transport and Mixing Processes in the Ocean by Dynamical Systems Techniques, Christopher Jones and Lawrence Pratt
9:45 a.m.	How a Physicist and a Mathematician Got Together and Did Something Useful in Brain Imaging—Lawrence Shepp and Jay Stein
10:45 a.m.	*Break*
11:00 a.m.	Panel Discussion: Multidisciplinary Research and Training in the Mathematical Sciences: Successes and Failures—Michael Tabor, Avner Friedman, Alan Newell, Nancy Sung, and Mary Wheeler
12:45 p.m.	*Lunch*
1:45 p.m.	Blue Lasers: Materials Growth, Characterization, and Computational Physics—David Bour and Chris Van de Walle
2:45 p.m.	Coping with Complex Surfaces: An Interface between Mathematics and Condensed Matter Physics—Jack Douglas and Fern Hunt
3:45 p.m.	*Break*
4:00 p.m.	Numerical Simulation of Subsurface Flow and Reactive Transport—Todd Arbogast and Mary F. Wheeler

---

NOTE: The workshop "Exploring the Interface Between Other Sciences and the Mathematical Sciences" was held March 25-26, 1999, at the National Academy of Sciences, Washington, D.C.

5:00 p.m.	Wavelets: A Synthesis of Ideas in Harmonic Analysis and Subband Filtering That Happened Serendipitously—Ingrid Daubechies and Martin Vetterli
6:00 p.m.	General discussion
6:15 p.m	*Adjourn for day*

March 26, 1998

8:15 a.m.	Language and Dynamical Systems: A View from the Bridge—Robert Berwick and Partha Niyogi
9:15 a.m.	Protein Folding Class Predictions—Temple Smith and James White
10:15 a.m.	Economics in Infinite Dimensional Spaces—Robert Anderson[1] and William Zame
11:15 a.m.	*Break*
11:30 a.m.	Roundtable Discussion: What Helps and What Hinders Collaboration between Fields in Academics and Industry? James Phillips, Suzanne Withers, and Margaret Wright
12:45 p.m.	Closing remarks
1:00 p.m.	*Adjourn*

SUMMARY OF PRESENTATIONS AND DISCUSSIONS

Chairman's Welcome
Thomas Budinger

After a general welcome and introductions, the Chair of the Committee on Strengthening the Linkages Between the Sciences and the Mathematical Sciences, Thomas Budinger, explained that the purpose of the workshop was not so much to look at interesting problems in science and mathematics but to consider how successful collaborations between science and mathematics came about, what factors contributed to their success, and what factors inhibited them. He thanked the presenters for their contributions and opened the workshop presentations.

---

[1] Anderson was unable to attend due to illness.

Introductory Remarks
Phillip Griffiths, Institute for Advanced Study

Griffiths described the genesis of this study in his own experiences. As provost at Duke University, he was struck when reviewing tenure files by the amount of mathematics being used throughout the university—not just in the physics department, but in unexpected places like the business school, the medical center, the engineering school, and the environment school. It seemed a sizable fraction of the university faculty were using mathematics in a substantive form in their research, but there was a low level of interaction between the mathematics department and these other departments. When he was chairing the NRC's Board on Mathematical Sciences, the board produced several reports on the interface between mathematics and specific areas of science, such as medical imaging. In reviewing those reports, he noted both the great potential use of mathematics in interesting scientific problems and the large number of interesting mathematics problems that were arising out of such applications. These two experiences were important factors in his early discussions with NRC staff to scope this study.

Griffiths also noted that science can now be considered to have three components: experimentation/observation, theory, and modeling/simulation. Each feeds into the other, and two of the three components clearly involve mathematics.

Finally, Griffiths expressed his hopes for the outcome of this study. He hoped to see a final report that included substantive accounts of interactions between mathematics and the other sciences and that drew from those accounts implications for the mathematics and other scientific communities.

The Elucidation and Quantification of Transport and Mixing Processes in the Ocean by Dynamical Systems Techniques
Lawrence Pratt, Woods Hole Oceanographic Institute
Christopher Jones, Brown University

Pratt, a physical oceanographer, and Jones, a mathematician, worked together on mixing and exchange in ocean current systems. The questions in oceanography included how mixing and exchange occur in major current systems such as the Gulf Stream and how current trajectories reduce to the physical features of the current. The resulting mathematical problems concerned how to understand dynamical systems with insufficient data to perform statistical analysis. The approach taken was to represent the velocity distribution in the current by a model, which led to an inverse problem: How much Lagrangian dynamics is needed to understand the entire velocity field? Results included a student thesis on Melnikoff functions and other mathematical spin-offs.

Pratt and Jones met when the Office of Naval Research (ONR) program officer funding Jones recognized the connection between his mathematics and problems in physical oceanography. Jones visited Woods Hole and was introduced to the sort of data on currents collected by physical oceanographers and the sort of dynamical systems they wished to understand. This first meeting led to an informal workshop, held at an inn on the coast of Rhode Island. In this relaxed setting, mathematicians and oceanographers met, with the goal of setting an agenda for collaboration. The motivation was a potential new ONR funding program for such work. This workshop did not result in any concrete work—no papers, no reports—but is remembered by Jones as a key event in the collaboration. Although no formal agenda for collaboration resulted, the researchers began to understand each other's languages, and the mathematicians began to formulate "big pictures" of what the main issues were in understanding the dynamics of ocean mixing and exchange.

Despite the fact that the ONR funding program had not yet materialized, the mathematicians got down to work. They were hampered by the lack of appropriate data on currents and by the lack of an explicit analytical expression for the velocity field. Success required that the mathematicians become accustomed to dealing in the imperfect world of physical oceanographers. As worked geared up, a postdoctoral fellow in mathematics was assigned to the project full-time, with funding from ONR. This provided the intense engagement necessary to really move the project out of its initial stages.

It took another meeting at Woods Hole before the concrete research agenda began to form and the really interesting mathematics inherent in these oceanographic problems began to become clear. Regular meetings between the oceanographers and mathematicians at various venues furthered the work. It took a period of several years for the work to reach this point.

In summarizing their observations about the collaboration, Jones and Pratt made several points. Jones noted that such interdisciplinary collaborations bring tremendous excitement to the field of mathematics, enriching mathematical investigations by the new problems that arise. The mathematicians involved needed to be patient and willing to work on two levels—first at the level of the application and only later at the second level of genuinely new and interesting mathematical problems. Many events without concrete results were, in retrospect, very critical, as they were part of a larger process that led to successful communication and collaboration between the researchers. Finally, Jones and Pratt raised the issue of incentives for collaborations. They felt that the long time needed to begin such a collaboration, a period that might not produce publications or other easily recognized results, was a disincentive, particularly for researchers in the early stages of their career. Research results were also not always readily published in journals relevant to both disciplines—not many mathematicians read the *Journal of Physical Oceanography*. Left open were some questions. Is it fair to get students or postdoctoral fellows involved in this area? Will they end up falling between two disciplines, with no job?

Technical Bibliography

Dutkiewicz, S. 1997. Intergyre Exchange: A Process Study. PhD dissertation, Graduate School of Oceanography, University of Rhode Island. 196 pp.

Flierl, G.R., P. Malanotte-Rizolli, and N.J. Zabusky. 1987. Nonlinear waves and coherent vortex structures in barotropic β-plane jets. J. Phys. Oceanogr. 17:1408-1438.

Pratt, L.J., M.S. Lozier, and N. Beliakova. 1995. Parcel trajectories in quasigeostrophic jets: Neutral modes. J. Phys. Oceanogr. 25:1451-1466.

Ridgway, K.R., and J.S. Godfrey. 1994. Mass and heat budgets in the East Australian current: A direct approach. J. Geophys. Res. 99(C2):3231-3248.

How a Physicist and a Mathematician Got Together and Did Something Useful in Brain Imaging
Lawrence Shepp, Rutgers University
Jay Stein, Hologic, Inc.

Shepp, a mathematician, and Stein, a physicist, collaborated to invent what was then known as the fourth-generation CAT scanner—a practical scanner with 600 photomultiplier detectors. The basic principle behind what is now known as CT scanning is the collection of simple X-ray images of the body from many different angles. Mathematical techniques allow one to reconstruct from these many images a single, more detailed image providing information on both bone structure and soft tissue. The basic problem facing Shepp and Stein was the artifacts generated by a system of many imperfectly balanced, nonidentical X-ray detectors. What artifacts would be generated? Stein presented a brief overview of the science and mathematics employed in their work. Using mathematics to come to an understanding of these artifacts, they were able to design a practical system and enable CT scanning with better resolution of soft tissue.

Shepp raised the question of why a mathematician would want to get involved in an applied problem. In his own case, he was motivated when his child was diagnosed with a brain tumor. It took such a great personal motivation for him to turn from more abstract work to this applied problem. He described pressure from mentors in mathematics not to work on medical applications—they felt it was a mistake that would damage his career. Shepp attributes such attitudes to a phenomenon he refers to as tribal bonding—people banding together in groups for purposes of security. Mathematics, like all the disciplines, suffers from such bonding, in his view, which makes it difficult for researchers, especially young researchers, to move into interdisciplinary collaborations.

Shepp noted, however, the great rewards of interdisciplinary collaborations: "If you take something head on, like Fermat's last theorem, you have very little chance of success and it takes enormous effort. If you work in an interdisciplinary field where nobody has been working before and bringing your own methods from mathematics, you are in a tremendous position. It is really easy to make a big splash." Shepp felt his recognition, despite the warnings of his colleagues, was a result of rather than in spite of his interdisciplinary work. He described this work as "the high point in . . . my scientific life, and no question about it. This interaction was really fantastic."

Technical Bibliography

Shepp, L., and J.A. Stein. 1977. Simulated artifacts in computerized tomography, Reconstructive Tomography in Diagnostics Radiology and Nuclear Medicine, M.M. Ter-Pogossian, ed. University Park Press, Baltimore, Md.

Shepp, L., J.A. Stein, and B.F. Logan. 1977. A variational problem for random young tableaux. Adv. Math. 26:206-222.

Panel Discussion: Multidisciplinary Research and Training in the Mathematical Sciences: Success and Failures
Michael Tabor, University of Arizona
Avner Friedman, University of Minnesota
Alan Newell, University of Warwick
Nancy Sung, Burroughs Wellcome Fund
Mary F. Wheeler, University of Texas at Austin

Tabor opened by speaking briefly of some of the observations he has made while developing interdisciplinary applied mathematics programs. First, he cautioned against what he termed fraud—repackaging existing research programs under the rubric "interdisciplinary" without actually developing a significant interdisciplinary interaction. Those interactions, he noted, took time and commitment to nurture. A program in biomathematics he helped organize took three years before solid enough relationships existed between the biologists and mathematicians to allow meaningful collaborations to begin—and in those 3 years, there were very few tangible benefits to be shown along the way. On the question of whether young researchers should be encouraged to engage in interdisciplinary pursuits, he noted that his current program does engage graduate students successfully in interdisciplinary work. He noted that such students require good dual mentors from both disciplines in order to obtain sufficiently solid training in both disciplines, but that such dual mentoring brings faculty partners from different departments into closer, more fruitful relationships. For mathematicians, he felt it was important to engage in what he termed "service mathematics"—assisting scientists with problems that involve little or no new development of mathematics and thus do not further the mathematician's own research career—in order to open doors to other disciplines. This was a means of testing the waters and finding out what the open research questions were in other fields that might be of interest to mathematics. Finally, he raised the difficulties inherent in the review of interdisciplinary proposals and questioned whether mechanisms currently exist at funding agencies to properly review truly interdisciplinary work.

Friedman discussed his experiences establishing and running the Minnesota Center for Industrial Mathematics. He saw the purposes of the center as threefold: (1) education, which in practice means courses for students, (2) research, which is carried out by students with mentors from both the university and industry, and (3) training, which students obtain through industrial internships. The center was born when the mathematics department at the University of Minnesota garnered the support of university administrators for additional faculty slots for such a center. The department provided two faculty slots for every one created by the university. As Friedman described it, obtaining administration support for a center for industrial mathematics was straightforward, but his colleagues in the mathematics department required more convincing, especially when the proposal for a PhD program was made. Two points were needed to convince the mathematics faculty. First, they had to be convinced that the PhD program would be built on solid, traditional mathematical training that did not cut corners to accommodate the need to spend time in applications. Second, supporters of the center raised the argument that to fail to embrace applications and interdisciplinary collaborations would eventually erode the program as the number of graduate students declined below a critical level. These two points eventually won the day, and Friedman reported that graduates of the center's programs are very successful in obtaining interesting, competitive industrial positions. Friedman noted three essential elements in starting up such a program. First, the individual(s) pursuing the program must be committed, as establishing an interdisciplinary program or collaboration requires significant time and energy. Second, resources in the form of time release for faculty are necessary. Finally, it is essential to success that any such program have the strong support of the chair of the mathematics department.

Sung described programs the Burroughs Wellcome Fund supports to bring scientists and mathematicians from other disciplines into problems with biological applications. In her experience, the Fund has two effective ways of bringing about interdisciplinary collaborations. The first is to encourage direct interactions between biologists and researchers from other disciplines. Institutions supported by the fund often use the mechanism of consortia to bring researchers together. The second method is to fund postdoctoral researchers or graduate students working at the research interface—these young researchers serve as a link between their faculty mentors and thus as a link between two fields. The fund is attempting to measure its success in breaking down research barriers by tracking the careers of the individuals it funds—what sort of jobs are they taking, what research are they publishing, and what journals are they publishing in? The programs were as yet too new to have definitive answers to these questions.

Newell spoke from his experience setting up interdisciplinary mathematics programs at Arizona and Warwick. He began by pointing out that strong disciplinary departments were necessary to nucleate good interdisciplinary centers. Researchers then had to be persuaded to leave the comfort of their own discipline's culture to interact with another discipline. He noted that the challenges of such interactions attracted high-quality students and, in his experience, raised the quality of students applying to the mathematics program. To support those students, it was necessary to establish an environment to encourage and nurture interactions. At Warwick, this was achieved by setting aside one afternoon a week for interdisciplinary speakers and seminars. Resources were also important—extra time was needed to establish solid interdisciplinary work. Thus administration support was also key, but generally easy to obtain as universities are dependent on their mathematics departments to provide training to large numbers of undergraduates from many departments. By taking responsibility for meeting undergraduate educational needs in mathematics—including any remedial work necessary for entering students—departments can raise their value to the university and be in an even stronger bargaining position with the administration of resources.

Wheeler described another model for interdisciplinary work used at the Texas Institute for Computational and Applied Mathematics (TICAM). Faculty involved in TICAM have appointments in their home departments and are affiliated with the institute. The institute involves the departments of mathematics, computer science, chemistry, physics, geology, astronomy, biology, and engineering. It has very strong ties with local industrial concerns. Students work in groups on applied problems identified by the industrial partners. The real-life problems motivate students and give them practical experience. Many have gone on to jobs at well-respected industrial firms such as Bell Labs, Shell, and Exxon.

The question-and-answer period revolved around whether it is best to provide students with interdisciplinary training or to provide them with solid disciplinary training and foster the traits—good communication, professional respect and curiosity, summed up by one participant as "entrepreneurial attitude"—that will enable them to use their training to take on interdisciplinary tasks. There was no consensus on this issue. Some participants' experience suggested that solid disciplinary training was important for both industrial and academic jobs, while others felt it should be possible to tolerate gaps in disciplinary training for students interested in interdisciplinary problems.

Blue Lasers: Materials Growth, Characterization, and Computational Physics
David Bour and Chris Van de Walle, Xerox Palo Alto Research Center

Bour and Van de Walle collaborate at Xerox PARC on the development of blue semiconductor lasers. These devices are sought because the shorter wavelength of blue light can provide higher storage density on DVD devices and, in combination with red and green semiconductor lasers already available, enable full-color LED displays. The problems facing Bour, the engineer, and Van de Walle, the theoretician, were essentially materials science problems. They needed to optimize the properties of the materials used in the semiconductors, then optimize their interfaces as they are layered in the semiconductor structure. The most efficient path to breakthroughs in these areas was a combination of experiment and modeling. An understanding of the physical phenomena underlying the materials properties, obtained through theory, guided the choice of materials and material growth mechanisms employed in experiments intended to yield a successful device.

The empirical approach, which is instinctive to engineers, was a barrier to Bour's acceptance of the contributions that Van de Walle could make to the project. Building trust took time and effort for the two researchers. They noted that it was necessary to learn how to present results to each other in a way that each could understand. They noted, however, that PARC has an environment that nurtures such interactions and that inherently values the contributions modeling can make to applications, and this encouraged their relationship.

Technical Bibliography

McCluskey, M.D., C.G. Van de Walle, C.P. Master, L.T. Romano, and N.M. Johnson. 1998. Large band gap bowing of $In_xGa_{1-x}N$ alloys. Appl. Phys. Lett. 72:2725.

Neugebauer, J., and C.G. Van de Walle. 1994. Atomic geometry and electronic structure of native defects in GaN. Phys. Rev. B 50:8067.

Van de Walle, C.G., and J. Neugebauer. 1997. Small valence-band offsets at GaN/InGaN heterojunctions. Appl. Phys. Lett. 70:2577.

Coping with Complex Surfaces: An Interface between Mathematics and Condensed Matter Physics
Jack Douglas and Fern Hunt, National Institute of Standards and Technology

Douglas and Hunt described their collaboration on problems of polymer dynamics and complex geometry of polymers. Douglas, a theoretical materials scientist, works in the area of phase transitions, polymers, membranes, and percolation theory—areas requiring the application of statistical physics. Douglas was motivated to meet and know his mathematical colleagues at the National Institute of Standards and Technology (NIST). As he put it, his research interest requires so many different areas of expertise—pattern formation problems, dynamics of phase separation, free-boundary problems such as dewetting—and one person can't know everything. As Hunt described it, Douglas "patrols the halls" and makes it a point to meet new researchers. He sought her out because aspects of their research interests overlap.

Hunt, a mathematician, has interest in the area of invariant measures and the ergodic theory of dynamical systems. Hunt's position at NIST calls for her to spend a significant portion of her time providing service mathematics to scientists and engineers at the Institute. She described her work with Douglas as very interesting and rewarding because it led to genuinely new mathematical problems for her to work on, in the area of boundary behavior of harmonic functions.

Hunt and Douglas both identified barriers to their collaboration. Hunt felt that there was increasing pressure at NIST to produce research results in the short term. This worked against establishing the rapport and relationship necessary to enable her collaboration with Douglas. Time pressures also came into play when it became apparent that some problems would be furthered only by large-scale computation—and there was insufficient time to devote to both the computations and the mathematics. Douglas posed as a barrier the difficulty in reading mathematical literature. He felt that mathematicians needed to spend more time developing the context of their research problems and explaining the motivating factor for pursuing the solution to a given mathematical problem.

On the positive side, Hunt noted that scientists at NIST were very supportive of mathematician colleagues and their research. This, she felt, contributed to a positive atmosphere for collaboration.

Technical Bibliography

Bernal, J., J.F. Douglas, and F.Y. Hunt. 1995. Probabilistic computation of Poiseville flow velocity fields. J. Math. Phys. 36:2386.

Numerical Simulation of Subsurface Flow and Reactive Transport
Todd Arbogast and Mary Wheeler, University of Texas at Austin

Arbogast (Department of Mathematics) and Wheeler (Departments of Aerospace, Engineering, Engineering Mechanics, and Petroleum and Geosystems Engineering) discussed ongoing work at the TICAM Center for Subsurface Modeling. The center uses high-performance parallel processing as a tool to model the behavior of fluids in permeable geologic formations such as petroleum and natural gas reservoirs, groundwater aquifers and aquitards, and shallow water bodies such as bays and estuaries. The research revolves around practical tasks such as contamination cleanup. In addition to the obviously challenging engineering problems such tasks pose, they also pose difficult mathematical and modeling problems. These problems involve highly nonlinear systems of equations that exhibit hysteresis. The equations exist on scales from the minute to the very large. Localized events provide singularities in the equations modeling the larger system. Large sets of data require challenging parallel computations.

The TICAM Center for Subsurface Modeling is composed of what Wheeler described as a close-knit team of faculty and research scientists. Expertise represented includes applied mathematics; engineering; physical, chemical, and geological sciences; and computer science. Participating students earn a degree in computational and applied mathematics. The center came about when the petroleum industry was downsizing and seeking to outsource much of its research. The university was able to capitalize on that, establishing an industrial affiliates program with the center. Dues for affiliation provided monies for foreign travel and other expenses not readily covered under federal research grants. Also, center investigators found that the industrial funding made it easier to attract the federal grants that provide the bulk of the center's resources.

Center researchers, both faculty and student, are very actively involved in interactions with the industrial affiliates. Workshops, meetings, poster sessions, speaker series, and affiliate visits to the center provide substantial opportunity for interaction. The industrial affiliates look to the center to provide the basic research needed to support practical applications.

Wheeler noted that the continuity of funding which the center had been able to achieve was important. This allowed research interactions to develop over time. The ability to support foreign travel and significant amounts of conference travel also contributed to the center's reputation and success. Finally, the intensive interactions with industry gave students training in presenting their work to a diverse audience.

Technical Bibliography

Arbogast, T., and M.F. Wheeler. 1995. A characteristics-mixed finite element method for advection dominated transport problems. SIAM J. Numer. Anal. 32:404-424.

Arbogast, T., M.F. Wheeler, and I. Yotov. 1996. Logically rectangular mixed methods for flow in irregular, heterogeneous domains. Pp. 621-628 in Computational Methods in Water Resources XI, Vol. 1, A. Aldama et al., eds. Computational Mechanics Publications, Southampton.

Arbogast, T., M.F. Wheeler, and N.-Y. Zhang. 1996. A nonlinear mixed finite element method for a degenerate parabolic equation arising in flow in porous media. SIAM J. Numer. Anal. 33:1669-1687.

Wavelets: A Synthesis of Ideas in Harmonic Analysis and Subband Filtering That Happened Serendipitously
Ingrid Daubechies, Princeton University
Martin Vetterli, University of California, Berkeley

Wavelet theory has its roots in harmonic analysis and the development of the Fourier transform. Wavelet expansions provide a better description of systems that contain phenomena occurring at multiple scales, from the very coarse to the very fine, than previously available algorithms. They have found widespread application in areas such as vision theory and signal compression. In particular, their application by electrical engineers to achieve better image compression produced new mathematical insights that led to new results in approximation theory. Vetterli described the development of wavelets as a two-way street in which insights and advances flowed back to both the mathematicians and the scientists involved in their development and use.

Daubechies and Vetterli outlined the history of wavelet theory. Its roots lie in several fields, and influence can be traced to harmonic analysis, standard coherent-state decompositions used in quantum mechanics, approximation theory, vision theory, and computer-aided geometric design. The synthesis of these roots into wavelet theory occurred over a relatively short period, 1982 to 1988, and involved encounters and interactions among a number of researchers from different fields. Daubechies traced it as follows.

A common friend introduced Alex Grossman, a theoretical physicist, to Jean Morlet, a geophysicist, who was working out new transforms that would allow better localization at high frequencies. Grossman used his expertise in coherent states to help Morlet put his work on a more sound mathematical basis, and the result was the first crude wavelet transform. Yves Meyer, a harmonic analyst, heard about Grossman and Morlet's work while standing in line for a photocopy machine and recognized that it was a rephrasing of earlier work by Calderon. Meyer contacted Grossman, and their work was advanced another step. Work by Meyer, Grossman, and Daubechies led Meyer to a new wavelet basis set, which the researchers would intuitively have thought couldn't exist. That basis was picked up by Stephane Mallat, a vision researcher, who heard about it from an old friend who was a graduate student of Meyer. The multilevel view prevalent in vision theory led to a different understanding of the basis construction, and a collaboration of Mallat and Meyer resulted in multiresolution analysis, a better tool for understanding most wavelet bases. Mallat had tried to relate the wavelet bases construction to practical algorithms that existed in electrical engineering. Using these algorithms as a point of departure, Daubechies developed a series of different bases constructions in which the wavelets were defined through the algorithm itself. For reasons he couldn't recall, Vetterli read Daubechies' paper, and his work with Daubechies to generalize the bases for use in signal compression led to yet more mathematical insights and fruitful research for both parties.

Daubechies and Vetterli emphasized several factors which they thought made the collaborations work. Serendipitous interactions triggered many of the key collaborations, but the wavelets problem provided topics of interest to both sides of the interaction, which kept all the contributors engaged. Openness on the part of the researchers—openness to looking at a problem from the perspective of another field, openness to talking to researchers in other fields, openness to those who were not expert in one's own field—was crucial. Daubechies, trained initially as a physicist, took possible applications seriously when writing up her results. She included a table of coefficients and a description of the algorithm so that the construction could be used even by those who could not or would not piece through the mathematical analysis in the remainder of the paper. She provided the reader with more context than usual in mathematics publications, trying to help the nonmathematical reader to understand the implications of her work. This was particularly important when Vetterli, the electrical engineer, read her paper—he commented that Daubechies had done an excellent job of explaining some rather esoteric points.

Vetterli commented that the U.S. tenure system could work against such interactions. He spent a year learning harmonic analysis and thus probably had fewer papers on his publication list when he came up for tenure review. His department chairman counseled him against writing a book on wavelets, since it would be only one line on his publication list. He also felt that, in general, the engineering and mathematics communities had different cultures, reward systems, and ways of approaching problems, which often made interactions difficult. Formal incentives and rewards for interdisciplinary work, he felt, would foster more interactions and successful research collaborations.

Technical Bibliography

Daubechies, I. 1988. Orthonormal bases of compactly supported wavelets. Commun. Pure Appl. Math. 41:906-966.

Daubechies, I. 1992. Ten Lectures on Wavelets. Philadelphia, Pa.: Society for Industrial and Applied Mathematics.

Vetterli, M., and C. Herley. 1992. Wavelets and filter banks: theory and design. IEEE Transactions on Signal Processing 40:2207-2232.

Vetterli, M., and J. Kovacevic. 1995. Wavelets and Subband Coding. Englewood Cliffs, N.J.: Prentice-Hall.

Language and Dynamical Systems: A View from the Bridge
Robert Berwick, Massachusetts Institute of Technology
Partha Niyogi, Bell Laboratories

Niyogi and Berwick worked together on problems of modeling the development of human language using dynamical systems. Such work is one approach in the attempt to understand, for example, how modern English developed from old English.

This work occurred under the auspices of the Center for Biological and Computational Learning at the Massachusetts Institute of Technology. This NSF-funded center had the aim of providing an explicit infrastructure for bringing together researchers from different disciplines. Graduate students and postdoctoral researchers at the center were required to have two advisors—one from mathematics or computer science and one from brain or cognitive science.

Berwick used the analogy of the coffeepot to describe the center. A coffeepot provides a nucleation site, where people meet and bump into each other, begin to talk, and learn from each other. The center performed such a function for these two research communities at MIT. It enabled young researchers like Niyogi to pursue an interdisciplinary interest for which they might not otherwise find support and mentoring.

Technical Bibliography

Niyogi, P., and R.C. Berwick. 1996. A language learning model for finite parameter spaces. Cognition 61:161-193.

Brent, Michael R., ed. 1997. Computational Approaches to Language Acquisition. Cambridge Mass.: MIT Press.

Protein Folding Class Predictions
Temple Smith, Boston University
James White, TASC, Inc.

Smith and White collaborated on research that sought to model protein folding using Markov models. Smith trained as a physicist and is now part of the Molecular Engineering and Research Center (MERC) at Boston University. White, an engineer, described his training as having a strong mathematical component. They felt that in their collaboration, Smith functioned as a molecular biologist and White as a mathematician.

They agreed that the value of their collaboration was that each partner brought a different skill and perspective to the task, which changed the way each thought about the research problem. Their collaboration was aided, they felt, by the relative similarity of their training—they felt they had more language in common than two researchers actually trained in molecular biology and mathematics would have had. White's company encouraged its employees to have contact with academic research centers. MERC—with its multidisciplinary faculty and graduate students drawn from computer science, chemistry, physics, electrical engineering, biology, and medicine—provided a venue for their collaboration to unfold. White and Smith were joint mentors for students at the center.

Smith commented on the difficulty of obtaining funding for interdisciplinary research. He noted, however, that while 20 years ago his proposals elicited questions about whether physics had anything to contribute to biology, that attitude is changing.

Technical Bibliography

White, J.V., C.M. Stultz, and T.F. Smith. 1994. Protein classification by stochastic modeling and optimal filtering of amino-acid sequences. Math. Biosc. 119:35-75.

Economics in Infinite Dimensional Spaces
William Zame, University of California, Los Angeles
Robert Anderson, University of California, Berkeley[2]

Anderson and Zame have collaborated on problems of commodities trading. The problems they have focused on concern continuous trading over a time period, which resolves to a problem involving an infinite number of commodities. Anderson trained as a mathematician, but his involvement in economics problems began as early as his graduate thesis. Zame also trained as a mathematician but came to economics much later in his career, when he had already held several faculty positions.

Zame came to economics through a personal contact. Near the end of a sabbatical leave at the University of California, Los Angeles, a colleague introduced him to problems in economics by taking him to the university's economics department and to a meeting of the econometrics society. He made another contact there, whom he later looked up when attending a mathematics conference in her home city. Over dinner, she described the problems she was

---

[2] Zame presented for both himself and Anderson, as Anderson was unable to attend due to illness.

working on. The conversation became a minitutorial in economics, and by the end Zame had found interesting problems to work on.

From there his training in economics was furthered by formal programs. He spent 18 months at an NSF-funded program in mathematical economics at the Institute for Mathematics and its Applications, in Minnesota. Next, he spent time at the Mathematical Sciences Research Institute in a similar program. Soon he was receiving job offers from economics departments.

Zame identified several factors that enabled his transition to mathematical economics. One was time—he was able to spend several sabbaticals in economics departments, interacting and learning from colleagues there. Another was good mentoring. He felt he had received much patient mentoring from economists as he learned the ropes and made mistakes—such as "solving" a problem with a mathematical model that allowed negative consumption.

Zame felt that the typical academic career path in economics hampered the movement of mathematicians into faculty positions in economics. Economics departments generally do not offer postdoctoral research positions. If a department is interested in a promising mathematician, it typically must offer him or her a tenure-track position—meaning the department has committed to the individual for 6 years, rather than the 2- or 3-year period typical for a postdoctoral fellow. Zame felt many departments were unwilling to commit such a large amount of resources to a person who might or might not make the transition to a new discipline successfully. To foster more math/economics interdisciplinary research, Zame felt that investments should be made in means to allow mathematicians and economists to spend time together—such as conferences, seminars, and sabbatical programs.

Technical Bibliography

Anderson, R.M., and W.R. Zame. 1997. Edgeworth's conjecture with infinitely many commodities. Econometrica 65:225-274.

Roundtable: What Helps and What Hinders Collaboration Between Fields in Academics and Industry?
James Phillips, The Boeing Company
Suzanne Withers, University of Washington
Margaret Wright, Bell Laboratories

The discussion identified several themes that had come from the various presentations:

- Time and patience were needed to develop individual interdisciplinary collaborations—to develop trust and rapport between collaborators, to understand each other's language sufficiently, to recognize the emerging questions underlying each research problem. Time is also needed to build up a culture more supportive of interdisciplinary interactions generally. Sometimes many small steps are required to see a result.
- Fruitful interdisciplinary interactions involved intellectual challenges in both disciplines. This was important to motivate both collaborators to invest the extra time and effort needed to undertake a collaboration.

- Mutual respect between collaborators and the ability to communicate problems and results were also key. Researchers often have stereotypical views about other disciplines that may prevent them from listening carefully to their colleagues in other fields, or even from listening at all. Differences in jargon between the disciplines can also make clear communication more difficult, and both parties must be willing to work to overcome them.
- Proper reward systems can make interdisciplinary collaborations easier. Industry generally has less trouble rewarding such work, provided it furthers company goals.

Agreement on the need for formal training across disciplines varied. Some participants felt it was sufficient to learn enough language of the other discipline to communicate; others advocated more formal course work across disciplines. All agreed that good mentoring was necessary for both young and well-established researchers to make the move to interdisciplinary work. All agreed that face-to-face interactions were essential—one cannot become an interdisciplinary researcher by just reading up on another field.

The question, Why encourage interdisciplinary interactions? was raised and then answered. Interdisciplinary research can be uniquely innovative and ground-breaking, leading to new approaches to research questions that would not have been generated in a monodisciplinary setting. The question, Why encourage math-science interactions over other interdiscipilinary efforts? was also raised and answered. Mathematics brings qualities such as rigor, abstraction, and generality to a problem. It has the ability to transcend an application. It gives structure to a problem and can bring a unique way of understanding to an application.

All the participants agreed that interdisciplinary research can be difficult, time-consuming, and frustrating. They also agreed that it can provide some of the most rewarding and challenging opportunities of a research career.

C

Partial Chronology of Previous Efforts to Strengthen Mathematics and Cross-Disciplinary Research

1981   Research Institutes
The Mathematical Sciences Research Institute at the University of California-Berkeley and the Institute for Mathematics and Its Applications at the University of Minnesota are created with NSF support, intensifying NSF's emphasis on infrastructure for mathematical research and opportunities for mathematics to influence scientific disciplines and industrial activities.

1981   The David Committee
The NRC establishes a committee, chaired by Edward E. David, Jr., of scientists and engineers to review the health of and support for research in the mathematical sciences in the United States.

1983   The Joint Policy Board for Mathematics
The American Mathematical Society, the Mathematical Association of America, and the Society for Industrial and Applied Mathematics (SIAM) create a nine-member joint executive action arm, the Joint Policy Board for Mathematics (JPBM), to begin implementing the recommendations of the David Committee. The JPBM emphasizes unity across the discipline.

1984   Renewing U.S. Mathematics: Critical Resource for the Future (NRC, 1984)
The 1984 David Report highlights the development of mathematics and its uses since World War II and calls attention to the following serious signs of trouble: (1) an impending shortage of U.S. mathematicians and (2) a marked imbalance between federal support of mathematical sciences research and support for related fields of science and engineering. Based on a careful analysis, it calls for more than doubling the FY 1984 federal support level for mathematical sciences and lays out a 10-year implementation plan, with specific roles for government, universities, and the mathematical sciences community. The report recognizes that the research community for mathematical sciences has changed in two important ways: (1) common research endeavors have blurred the boundaries of the major disciplines and (2) mathematics is increasingly looking outward, toward its interaction with science and technology. The report specifically recommends that ". . . departments should give increased recognition to faculty . . . who interact with collaborators from other disciplines."

1984   The Board on Mathematical Sciences
In December 1984, the NRC establishes the Board on Mathematical Sciences (BMS). It does this to provide a focus for concern at the NRC about issues affecting the mathematical sciences, to provide objective advice to federal agencies, and to identify promising areas of mathematics research along with suggested mechanisms for pursuing them.

NOTE: Entries to 1990 draw largely on the list in NRC (1990), pp. 33-36.

1985   The Mathematical Sciences Education Board
At the urging of the mathematical sciences community, the NRC establishes the Mathematical Sciences Education Board (MSEB) to provide "a continuing national assessment capability for mathematics education" from kindergarten through college. A 34-member board is appointed that is a unique working coalition of classroom teachers, college and university mathematicians, mathematics supervisors and administrators, members of school boards and parent organizations, and representatives of business and industry. This step reflects another of the basic recommendations of the David Committee: strong involvement of all sectors of the mathematical sciences community in issues of precollege education.

1986   Board on Mathematical Sciences' Science and Technology Week Symposium
This was the first of a series of annual symposia highlighting the role of mathematical sciences research in the sciences and engineering for an audience of scientists and policy makers.

1987   Project MS 2000
At the urging of JPBM and under the supervision of the BMS and MSEB, the NRC launches a comprehensive review of the college and university mathematics enterprise through the Mathematical Sciences in the Year 2000 (MS 2000) project.

1987   Science and Technology Centers: Principles and Guidelines. A Report by the Panel on Science and Technology Centers (NRC, 1987)
This NRC panel recommends that the primary goal of a proposed program of science and technology centers should be to exploit science where the complexity of the research problems or the resources needed to solve them require the advantages of scale and duration, or where facilities can be provided only by a centralized mode of research. The principal criterion used for evaluating proposals would be the scientific quality of the research. The panel cautions that interdisciplinary research, although essential for the solution of many problems, should be pursued only when there is a demonstrated need or opportunity. An initiative is launched by NSF to support important basic research and education activities and to encourage technology transfer and innovative approaches to interdisciplinary problems, by establishing centers devoted to critical areas of science and technology.

1988   Cross-Disciplinary Research in the Statistical Sciences: Report of a Panel of the Institute of Mathematical Statistics (IMS, 1988)
An influential report assesses the status of cross-disciplinary statistical research and makes recommendations for the future. Two principles endorsed by the panel are that (1) advances in substantive knowledge and in statistical theory are virtually inseparable and (2) the continued health of statistics strongly depends on research stimulated by and directed at problems in many other disciplines. Believing that "a continuing effort is required to monitor the health of cross-disciplinary statistical research," the panel recommended that the NRC, through its Committee on Applied and Theoretical Statistics and its Committee on National Statistics, undertake this effort. Finding that "constrained resources and the existing infrastructure within the government, academia, and industry thwart the growth and development of needed cross-disciplinary research," the panel recommended that an institute for statistical sciences be established—a recommendation that led to the establishment of the National Institute of Statistical Science.

1988   Removing the Boundaries: Perspectives on Cross-Disciplinary Research (Sigma Xi, 1988)
This Sigma Xi report explores the issue of multidisciplinary science and mathematical science, describing the trend toward boundary-breaking research.

1990   Renewing U.S. Mathematics: A Plan for the 1990s (NRC, 1990)
This report, which is a 5-year update of the 1984 David Report, describes emerging research opportunities and new challenges for government, universities, and the mathematical sciences community to continue to renew U.S. mathematics. It notes that collaborating with other disciplines will help the mathematical sciences become increasingly robust and valuable.

1990   Interdisciplinary Research: Promoting Collaboration Between the Life Sciences and Medicine and the Physical Sciences and Engineering (NRC/IOM, 1990)
Addressing the growing and urgent need to create mechanisms within universities, government, and industry to facilitate a flow of knowledge and researchers across the interfaces of the physical/engineering/mathematical sciences and the life/medical sciences, this IOM/NRC report endorses actions to reduce or remove obstacles to fruitful collaborative research across traditional disciplines. The report targets six critical elements in promoting research collaboration: (1) administration and institutional support, (2) availability of adequate funding, (3) open communication and collegiality, (4) overlapping educational experience, (5) availability of collaborators, and (6) opportunities for practical application and technology transfer.

1991   Mathematical Sciences, Technology, and Economic Competitiveness (NRC, 1991a)
Addressed to members of the mathematics community, corporate decision makers, and policy makers, this NRC report documents the importance of quantitative reasoning, supported by computational and mathematical models, to all aspects of the complete product cycle and to the economic competitiveness of U.S. industry.

1991   Mathematical Foundations of High-Performance Computing and Communications (NRC, 1991b)
This NRC report examines the elements of the federal government's high-performance computing and communications (HPCC) program and explicitly identifies the role of the mathematical sciences community in that effort and in the "grand challenges" of computational science. Two primary conclusions are that (1) the goals of the HPCC program cannot be met—and progress in solving grand challenge problems cannot continue—without active involvement of the mathematical sciences research community and (2) research and education relative to HPCC and grand challenge problems must link a broad range of sciences, the mathematical sciences, and computational sciences. Multidisciplinary settings offer the best chance for success.

1993   Mathematical Research in Materials Science: Opportunities and Perspectives (NRC, 1993)
This NRC report identifies research opportunities at the interface between materials science and the mathematical sciences and recommends ways of overcoming cultural and other obstacles to pursuing such research.

1993   Science, Technology, and the Federal Government: National Goals for a New Era (NAS, NAE, IOM, 1993)
The NAS Committee on Science, Engineering, and Public Policy report proposes a new covenant between science, technology, and society, which includes the condition that the funders of research allow interdisciplinary research to succeed by removing barriers to emerging areas of research and by encouraging institutional structures that enable the flow of interdisciplinary opportunities.

1994   Recognition and Rewards in the Mathematical Sciences (AMS, 1994)
This JPBM report finds that the current reward structure tends to discourage researchers who wish to cross-disciplinary boundaries. The report encourages the mathematical sciences community to value a number of activities, including outreach activities and interdisciplinary pursuits.

1994   Modern Interdisciplinary University Statistics Education: Proceedings of a Symposium (NRC, 1994)
The NRC Committee on Applied and Theoretical Statistics publishes this collection of discussions and presentations from a 1993 symposium to initiate a process of long-overdue change in upper-undergraduate, graduate, and postdoctoral education for statisticians and to stimulate the incorporation of interdisciplinary experience and realistic apprenticing in the nation's programs for statistical science majors, advanced-degree candidates, and postdoctoral students.

1995   The SIAM Report on Mathematics in Industry (SIAM, 1995)
The first phase of the Mathematics in Industry study, sponsored by NSF and the National Security Agency, characterizes the working environment of nonacademic mathematical sciences; summarizes the views of nonacademic mathematical scientists and managers on the skills needed for success and the training provided by a traditional graduate education; and suggests strategies for enhancing graduate education in mathematical sciences, nonacademic career opportunities, and application of mathematical sciences to nonacademic environments. Acknowledging the overwhelming interdisciplinary nature of the nonacademic research environment, the report identifies important traits in nonacademic mathematical scientists, which include interest in, knowledge of, and flexibility across applications; communication skills; and adeptness at working with colleagues, including those having expertise outside mathematical sciences.

1995   Mathematical Challenges from Theoretical/Computational Chemistry (NRC, 1995)
This NRC report finds that the needs of the theoretical/computational chemistry community create opportunities for synergistic research with almost the entire mathematical sciences community. The report describes prior fruitful collaborations, identifies some areas for potential interaction, and finds that active encouragement of further collaborations is likely to accelerate research progress.

1996   Modeling Biological Systems: A Workshop (NSF, 1996)
This NSF-funded report of a workshop discusses areas of opportunity for future research and means by which enhanced opportunities for cross-disciplinary research and training involving

biological and mathematical scientists can be promoted. The report complements the more exhaustive report Mathematics and Biology: The Interface—Challenges and Opportunities, published by NSF in 1992.

1997 Preserving Strength While Meeting Challenges: Summary Report of a Workshop on Actions for the Mathematical Sciences (NRC, 1997)
This NRC report highlights critical issues, including the need to enhance connections with other disciplines and rethink faculty evaluation in ways that may include adopting tenure and promotion criteria that reward interdisciplinary efforts.

1997 International Benchmarking of U.S. Mathematics Research (NAS, NAE, IOM, 1997)
The NAS Committee on Science, Engineering, and Public Policy assessment finds that U.S. mathematics is thriving and that its ties with other sciences and engineering are growing and deepening, but notes several critical issues—including the increasing demand for interdisciplinary research—that need be taken seriously.

1997 Organizing for Research and Development in the 21st Century: An Integrated Perspective of Academic, Industrial, and Government Researchers (Eisenberger et al., 1997)
This NSF- and DOE-funded report identifies barriers to cross-disciplinary research and makes recommendations for alleviating them. Barriers to cross-disciplinary research include tenure difficulties for junior faculty and the organization of resources along departmental lines within universities. The report recommends that funding sources continue to invest in high-risk, high-payoff research, and since many of these opportunities lie within the scope of cross-disciplinary research, both funding agencies and departments will need to be flexible to accommodate them.

1998 Report of the Senior Assessment Panel of the International Assessment of the U.S. Mathematical Sciences (NSF, 1998)
This NSF-sponsored report finds the state of U.S. mathematics to be preeminent but fragile. Academic mathematics is insufficiently connected to mathematics outside the university or to other disciplines. The panel notes that the structure of universities and the narrow vision of mathematics within mathematics departments militates against multidisciplinary research. Given that scientific problems of the future will be extremely complex and will require collaborative mathematical modeling, simulation, and visualization, the report encourages funding agencies to provide financial support that recognizes and rewards multidisciplinary activities and to recognize the long time required to become competent in that work.

1998 Strengthening Health Research in America: Philanthropy's Role (American Cancer Society et al., 1998)
This report of a workshop sponsored by the American Cancer Society, the Burroughs Wellcome Fund, the Howard Hughes Medical Institute, and the Pew Charitable Trusts identifies new and important areas of research for foundations to sponsor alone, in partnership with each other, and with other research entities to creatively stimulate the entire health services research environment. Among the new and important areas of research are those at the interface between the sciences, such as between biology and social science or biology and mathematics. The report

claims that multidisciplinary research may be critical to the success of academic health centers, now experiencing financial strains.

1998 Unlocking Our Future: Toward a New National Science Policy (U.S. Congress, House, 1998)

Vice Chairman Ehlers' Report to the House Committee on Science recommends that Congress make stable and substantial federal funding for fundamental science research a high priority. Noting that the practice of science is becoming increasingly interdisciplinary and that progress in one discipline is often propelled by advances in other, seemingly unrelated fields, the report recommends that the federal government fund basic research in a broad spectrum of scientific disciplines, mathematical sciences, and engineering and that it resist concentrating funds in a particular discipline.

1998 DOE/NSF Initiative on Computational Science

DOE and NSF cosponsor seven workshops to establish the scientific basis of a multi-million-dollar initiative involving cross-disciplinary research. The main goal is to identify opportunities, benefits, common needs, and barriers for the computational techniques needed to advance areas of science from medicine to geophysics. The areas considered by national and international experts are applied mathematics and computer science techniques, data analysis and management, simulation and modeling, nonlinear complex phenomena, geochemistry large quantum mechanical systems, and materials and geophysics. The plan is to take the results of these workshops and shape them into an action plan that will maximize the impact of a national investment in applications of computers and associated mathematical science to the solution of major problems in science, medicine, and the environment.

1998 The Genetic Architecture of Complex Traits Workshop Report and Recommendations (NIH, 1998a) and New Approaches to the Study of Complex Biological Processes Workshop Report (NIH, 1998b)

The first of these two NIH workshops focuses on means of increasing the rate of progress and improving the quality of research on the analysis of complex traits. The second workshop, originally titled "Biological Systems Analysis," was designed to provide the National Institute of General Medical Sciences with a perspective on the emergence of new conceptual and experimental approaches to the study of complex processes such as genetic circuitry, metabolic regulation, and macromolecular assembly. The participants in these workshops recommended three classes of initiatives: (1) the support of cross-disciplinary research with the specific objective of attracting investigators trained in the mathematically based disciplines (physics, engineering, computer science, applied mathematics, and chemistry) to the study of biomedical problems, (2) the development of workshops and other vehicles to train established biomedical scientists in new, quantitative approaches to their fields of study and, reciprocally, to acquaint established, mathematically expert nonbiologists with biological problems, and (3) the promotion of interdisciplinary training for scientists and researchers at the pre- and postdoctoral levels. The resulting programs begun at NIH, such as Quantitative Approaches to Complex Biological Problems, are discussed in Appendix D.

1999 Towards Excellence: Leading a Mathematics Department in the 21st Century (AMS, 1999)
This AMS Task Force report concludes that to remain strong and maintain their quality, mathematics departments must, among other things, build strong relationships with other departments on campus. The Task Force found that mathematics departments generally had a reputation for being too insular, and deans viewed this as a problem for both the teaching and the research mission of the department and the university.

REFERENCES

American Cancer Society, Burroughs Wellcome Fund, Howard Hughes Medical Institute, and the Pew Charitable Trusts. 1998. Strengthening Health Research in America: Philanthropy's Role. Available at <http://www.pewtrusts.com/pubs/publications.cfm>.

American Mathematical Society (AMS), Joint Policy Board for Mathematics. 1994. Recognition and Rewards in the Mathematical Sciences. Washington, D.C.: AMS.

American Mathematical Society, Task Force on Excellence. 1999. Towards Excellence: Leading a Mathematics Department in the 21st Century. Washington, D.C.: AMS. Available at <http://www.ams.org/towardsexcellence/>.

Eisenberger, P.M., A.R. Faust, and M. Knotek, eds. 1997. Organizing for Research and Development in the 21st Century: An Integrated Perspective of Academic, Industrial, and Government Researchers. Princeton, N.J.: Princeton Materials Institute.

Institute of Mathematical Statistics (IMS). 1988. Cross-Disciplinary Research in the Statistical Sciences. Available at <http://www.niss.org/reports/crossd2.html>.

National Academy of Sciences (NAS), National Academy of Engineering (NAE), and Institute of Medicine (IOM), Committee on Science, Engineering, and Public Policy. 1993. Science, Technology, and the Federal Government: National Goals for a New Era. Washington, D.C.: National Academy Press.

National Academy of Sciences (NAS), National Academy of Engineering (NAE), and Institute of Medicine (IOM). 1997. International Benchmarking of U.S. Mathematics Research. Washington, D.C.: National Academy Press.

National Institutes of Health (NIH), National Institute of General Medical Sciences. 1998a. The Genetic Architecture of Complex Traits Workshop Report and Recommendations. Available at <http://www.nih.gov/nigms/news/reports/genetic_arch.html>.

National Institutes of Health, National Institute of General Medical Sciences. 1998b. New Approaches to the Study of Complex Biological Processes Workshop Report. Available at <http://www.nih.gov/nigms/news/reports/complexbio.html>.

National Research Council. 1984. Renewing U.S. Mathematics: Critical Resource for the Future. Washington, D.C.: National Academy Press

National Research Council. 1987. Science and Technology Centers: Principles and Guidelines. Washington, D.C.: National Academy Press.

National Research Council. 1990. Renewing U.S. Mathematics: A Plan for the 1990s. Washington, D.C.: National Academy Press.

National Research Council. 1991a. Mathematical Sciences, Technology, and Economic Competitiveness. Washington, D.C.: National Academy Press.

National Research Council. 1991b. Mathematical Foundations of High-Performance Computing and Communications. Washington, D.C.: National Academy Press.

National Research Council. 1993. Mathematical Research in Materials Science: Opportunities and Perspectives. Washington, D.C.: National Academy Press.

National Research Council. 1994. Modern Interdisciplinary University Statistics Education: Proceedings of a Symposium. Washington, D.C.: National Academy Press.

National Research Council. 1995. Mathematical Challenges from Theoretical/Computational Chemistry. Washington, D.C.: National Academy Press.

National Research Council. 1997. Preserving Strength While Meeting Challenges: Summary Report of a Workshop on Actions for the Mathematical Sciences. Washington, D.C.: National Academy Press.

National Research Council (NRC) and Institute of Medicine (IOM). 1990. Interdisciplinary Research: Promoting Collaboration Between the Life Sciences and Medicine and the Physical Sciences and Engineering. Washington, D.C.: National Academy Press.

National Science Foundation (NSF). 1996. Modeling Biological Systems: A Workshop. Available at <http://www.nsf.gov/bio/pubs/mobs/stmobs.htm>.

National Science Foundation (NSF). 1998. Report of the Senior Assessment Panel of the International Assessment of the U.S. Mathematical Sciences. NSF9895. Arlington, Va.: National Science Foundation.

Sigma Xi. 1988. Removing the Boundaries: Perspectives on Cross-Disciplinary Research. Research Triangle Park, N.C.: Sigma Xi Publications.

Society for Industrial and Applied Mathematics (SIAM). 1995. The SIAM Report on Mathematics in Industry. Philadelphia, Pa.: SIAM.

U.S. Congress, House of Representatives, Committee on Science. 1998. Unlocking Our Future: Toward a New National Science Policy. Available at <http://www.house.gov/science/science_policy_study.htm>.

D

Federal Agencies That Provide Funding Opportunities

The following material is a beginner's guide to some of the federal agencies that support crosscutting research, education, and training having both scientific and mathematical components. It is not intended to be an exhaustive list of federal agencies or their programs.

AGENCIES SUPPORTING GENERAL SCIENCE AND TECHNOLOGY

National Institutes of Health
<www.nih.gov>

Offices, Institutes, and Centers

The NIH consists of institutes and centers funded independently by Congress and centrally administered by the Office of the Director:

National Cancer Institute (NCI),
National Eye Institute (NEI),
National Human Genome Research Institute (NHGRI),
National Heart, Lung, and Blood Institute (NHLBI),
National Institute on Aging (NIA),
National Institute on Alcohol Abuse and Alcoholism (NIAAA),
National Institute of Allergy and Infectious Diseases (NIAID),
National Institute of Arthritis and Musculoskeletal and Skin Diseases (NIAMSD),
National Institute of Child Health and Human Development (NICHD),
National Institute on Deafness and Other Communication Disorders (NIDCD),
National Institute of Dental Research (NIDR),
National Institute of Diabetes and Digestive and Kidney Diseases (NIDDKD),
National Institute on Drug Abuse (NIDA),
National Institute of Environmental Health Sciences (NIEHS),
National Institute of General Medical Sciences (NIGMS),
National Institute of Mental Health (NIMH),
National Institute of Neurological Disorders and Stroke (NINDS),
National Institute of Nursing Research (NINR),
National Library of Medicine (NLM),
National Center for Research Resources (NCRR),
John E. Fogarty International Center (FIC),
Warren Grant Magnuson Clinical Center (CC),
Center for Information Technology (CIT), and
Center for Scientific Review (CSR, formerly DRG).

The research activities of the institutes and centers are tied to the Office of the Director through the Office of Extramural Research, which oversees extramural activities conducted through grants, contracts, and cooperative agreements, and to the Office of Intramural Research, which coordinates research that is conducted mainly on-site by NIH personnel. A central listing of extramural programs, which draw approximately 80 percent of the NIH budget, is available in the *NIH Guide for Grants and Contracts,* at <www.nih.gov/grants/guide/index.html>.

Support Mechanisms and Cross-Disciplinary Programs

Most off-site research is funded through the Office of Extramural Research. Funding mechanisms, which include center grants (e.g., P30), program project grants (P01), and small grants (R03), are described in *Research Grants,* at <www.nih.gov/grants/policy/emprograms/overview/resrchgr.htm>.

It may be difficult for a mathematical scientist without a proven track record at NIH to receive funding from it. Forming a collaboration with an established NIH scientist is a good way to obtain initial funding from NIH and establish oneself within the agency.[1] NIH sponsors programs that allow for—or specifically promote—such collaborations, either on- or off-site. Some of these programs and others that promote cross-disciplinary education and training[2] are listed below:

- *Genetic Basis of Complex Behaviors, PA-98-097,* solicits applications for multidisciplinary, methodologically rigorous programs of neuroscience research that will use advanced techniques for statistical and molecular genetic analysis in human and animal populations to elucidate the genetic basis of complex behaviors. Provides up to 5 years support.
- *Genetic Architecture of Complex Phenotypes, PA-98-078,* supports new studies on the architecture of complex phenotypes, including research using human and model systems as well as research using theoretical approaches. Studies targeted by this program are expected, in part, to increase the quantity and quality of population-based data, lead to development of mathematical and statistical tools for analyzing measured genotype data, and create biologically relevant models for understanding the origins, roles, and implications of genetic variation in causing variation in phenotypes. Duration of grants varies with type of support, which can be principal investigator- or program-based.
- *Supplements for the Study of Complex Biological Systems, PA-98-024,* supports new quantitative approaches to the study of complex, fundamental biological processes by encouraging nontraditional collaboration across disciplinary lines. Term of award is limited by the funding period of the parent grant.
- *Quantitative Approaches to the Analysis of Complex Biological Systems, PA-98-077,* supports research projects that develop quantitative approaches to describe, analyze, and predict the behavior of complex biological systems, especially those requiring the integration of potentially large amounts of molecular, biochemical, cell biological, and physiological data.

---

[1] Names of NIH principal investigators (for potential collaboration) can be found through the Community of Science database at <www.cos.com>. Similarly, a scientist can use this database to identify mathematical scientists funded through NIH or other federal agencies.
[2] Some, but not all, NIH research training opportunities can be found at <grants.nih.gov.training/>.

These projects are expected to require the participation of individuals with diverse expertise and therefore to be of a collaborative and cross-disciplinary nature. Applicants are strongly encouraged to consider research areas in which systems approaches are likely to make significant contributions. There is a particular interest in studies using mathematical, computational, or theoretical approaches to understanding the fundamental biological mechanisms underlying behavior/molecular genetic analysis in human and animal populations to elucidate the genetic basis of complex behaviors. Provides up to 5 years support.

- *Fellowships in Quantitative Biology, PA-98-082,* encourages highly qualified individuals with doctoral training in the "traditional quantitative disciplines (such as mathematics, physics, engineering, and computer science) and biology" to obtain additional training in biological areas congruent to the mission of NIGMS, the institute tasked with supporting basic biomedical research that is not targeted to specific diseases and disorders. Up to 3 years of support.

- *Short Courses on Mathematical and Statistical Tools for the Study of Complex Phenotypes and Complex Systems, PA-98-083,*[3] provides support for short courses or workshops to assist scientists in preparing for research on complex phenotypes and complex systems by obtaining a solid understanding of available mathematical and computational tools and the requisite instruction in mathematical languages and applications of mathematics and statistics in order to facilitate collaboration with mathematical scientists on biological complexity. Support is limited to highly focused courses that reach a wide audience of scientists; it is not intended for university course or curriculum development. Up to 5 years of support.

- *Curriculum Development Award in Genomic Research and Analysis, PAR-98-063,* supports the development of courses and curricula designed to train interdisciplinary scientists who combine a knowledge of genetics and genomics research with expertise in computer science, mathematics, chemistry, physics, engineering, or closely related sciences. A collaborator must be identified who will contribute to the interdisciplinary nature of the courses or curricula. From 3 to 5 years of support.

- *Epidemiologic Research on Drug Abuse, PA-99-002,* supports studies that focus on defining factors and patterns associated with the initiation, escalation, continuation, and cessation of drug use and on associated antisocial, health-threatening, and other problematic behaviors that arise as a result of drug abuse. Development and application of innovative sampling, surveillance, ethnographic, and data collection methods and refinement of statistical tools to analyze epidemiologic data will be encouraged.

- *The Investigator-Initiated Interactive Research Projects Grants, PA-96-001,* offers a means of promoting collaborative efforts between or among projects that are scientifically related, while providing a record of independently obtained awards and retaining the research autonomy of each principal investigator.

- *Medical Informatics Training Program,* at <www.lhncbc.nlm.nih.gov/mitp/>, supports visiting scientists and students for research participation at the National Library of Medicine's Lister Hill National Center for Biomedical Communications. Programs include the following:

—Research Participation Program in Imaging Applications, which offers fellowships to faculty (3 months of funding) and students (3 months to 1 year of funding) pursuing research in image processing, object-oriented databases, visualization, and Internet distribution networks and

---

[3] See addendum, Notice 98-12, <www.nih.gov/grants/guide/notice-files/not98-112.html>.

—Research Program in Medical Informatics for Visiting Faculty, which provides faculty and research staff members at accredited colleges, universities, and technical institutes the opportunity for short-term visits, summer visits, or sabbatical leaves at the enter to pursue collaborative research in a variety of areas of medical informatics.

An NIH researcher seeking funds to explore a high-risk, high-payoff area of cross-disciplinary research should consider applying for a short-term, R21 grant from NIGMH: *Exploratory Studies for High-Risk/High-Impact Research, PA-97-049* (at <www.nih.gov/nigms/funding/pa/r21.html>. If the initial research holds promise, the principal investigator can then apply for another grant to further the research, possibly through a cross-disciplinary collaboration.

Program Announcements

A list of all notices, program announcements, and requests for applications (RFAs) published through the Office of Extramural Research is available through the *NIH Guide for Grants and Contracts* at <www.nih.gov/grants/guide/index.htm> or <grants.nih.gov/grants/funding/welcomewagon.htm>. A specific program announcement, PA-YY-NNN, will be listed under the address <www.nih.gov/grants/guide/pa-files/PA-YY-NNN.html>.

Review

NIH employs a two-part review process involving (1) initial review groups (IRGs) and (2) institute/center-based advisory councils. There are approximately 20 IRGs, organized by subject matter and composed of more focused study sections, which generally consist of 18-20 experts. The advisory councils are composed of leaders in the basic sciences (and mathematical science), medical sciences, education, or public affairs, who are appointed for each institute or center by its director.

A request for a research grant submitted by one or more prospective principal investigators is initially sent to the Center for Scientific Review, where it is considered by a referral officer, who decides which IRG and which institute(s)/center(s) would be most suitable to fund the application. (The referral office considers requests for both study section and institute/center assignments.) The assignment process is a flexible one, with interaction available on a case-by-case basis among referral officers, study section scientific review administrators (SRAs), institute program representatives, and applicants.

As applications are assigned to study sections, the SRA decides which study section members are best suited to review the proposal or act as discussants. Typically, two or three members are assigned as reviewers and one or two others serve as discussants. Study sections meet between mid-February and mid-March to discuss the scientific and technical merit of proposals. As of October 1997, all unsolicited[4] research grant applications are reviewed according to five criteria established by the NIH Committee on Improving Peer Review:

1. **Significance.** Does this study address an important problem? If the aims of the application are achieved, how will scientific knowledge be advanced? What will be the effect of these studies on the concepts or methods that drive this field?

2. **Approach.** Are the conceptual framework, design, methods, and analyses adequately developed, well integrated, and appropriate to the aims of the project? Does the applicant acknowledge potential problem areas and consider alternative tactics?

3. **Innovation.** Does the project employ novel concepts, approaches, or methods? Are the aims original and innovative? Does the project challenge existing paradigms or develop new methodologies or technologies?

4. **Investigator.** Is the investigator appropriately trained and well suited to carry out this work? Is the work proposed appropriate to the expertise level of the proposed principal investigator and other researchers (if any)?

5. **Environment.** Does the scientific environment in which the work will be done contribute to the probability of success? Do the proposed experiments take advantage of unique features of the scientific environment or employ useful collaborative arrangements? Is there evidence of institutional support?

The products of the deliberations are an evaluative summary statement; a priority score for each application found to have significant and substantial merit; and a percentile score of each application (ranked against all applications reviewed by the same review group within the past 3 years). At this point, NIH's program officials become the applicant's link to the NIH with regard to the interpretation of reviews and disposition of the application.

Usually, an application proceeds to the second level of review if its scientific merit rating places it among the top two-thirds of all applications. This level of review involves the institute's or the center's advisory council, which meets in May/June of each year and considers both the proposal's scientific merit ratings and its importance to the mission of the institute or center. In some cases, NIH expands this phase of review to include site visits.

Tips

Applicants are generally encouraged to talk to program directors and other NIH contacts listed on announcements and solicitations, to consider applying for special programs, and to read the *NIH Guide for Grants and Contracts*. Mathematical scientists seeking initial support should take the following steps:

1. Contact NIH staff members in charge of IRGs having potential overlap with the health-related component of the proposed research and determine their willingness to convene a

---

[4]Proposals submitted in response to normal program announcements are considered unsolicited. Solicited grants include RFA program announcements with special receipt dates (PAR), Small Business Innovation Research (SBIR) grants, and Small Business Technology Transfer (STTR) grants. RFAs and other solicitations will contain specific criteria for scientific peer review.

noncommissioned special emphasis panel (SEP) having expertise in the appropriate mathematical science(s) to serve as an ad hoc study section during the initial step of the review process;

2. Specifically request the IRG and type of SEP most appropriate for the proposal; and

3. Clearly tie the proposed project to those areas of health-related research and training discussed in the program announcement/solicitation to which application is being made and, where possible, further relate the proposed project to the mission of the sponsoring institute or center(s).

The appeals process is discussed at <www.nih.gov/nigms/funding/appeals.html>.

National Science Foundation
<www.nsf.gov>

Offices and Directorates

The Office of the Director includes several staff offices, including the Office of Polar Programs and the Office of Integrative Activities. It oversees all NSF activities, from the development of policy priorities to the establishment of administrative and management guidelines, including long-range planning. The NSF is further organized into the following directorates:

Biological Sciences (BIO),
Computer and Information Science and Engineering (CISE),
Education and Human Resources (EHR),
Engineering (ENG),
Geosciences (GEO),
Mathematical and Physical Sciences (MPS), and
Social, Behavioral, and Economic Sciences (SBE).

Detailed information about the organizational structure of the NSF is available online at <www.nsf.gov/home/nsforg/orglist.htm>. A guide to grants may be found at <www.nsf.gov/pubs/1999/nsf992/start.htm.

Crosscutting Programs

Many, but not all, crosscutting programs are listed on the Web sites for crosscutting programs, <www.nsf.gov/home/crssprgm/start.htm>, and for the Division of Mathematical Sciences within MPS, <www.nsf.gov/mps/dms/start.htm>.

Support Mechanisms for Crosscutting Research Involving Mathematics

The funding mechanisms at NSF are flexible, offering a variety of means by which crosscutting research involving mathematics can be supported: within a single directorate; between directorates; and among the NSF and other organizations, including other federal agencies, international agencies, and private organizations. Specific programs that support cross-disciplinary collaborations between scientists and mathematical scientists include the following:

- *Grant Opportunities for Academic Liaison with Industry (GOALI), NSF-98-142,* supports an eclectic mix of industry-university linkages. Special interest is focused on affording opportunities for (1) faculty, postdoctoral fellows, and students to conduct research and gain experience with production processes in an industrial setting; (2) industry scientists, mathematical scientists, and engineers to bring industry's perspective and integrative skills to academe, and (3) interdisciplinary university-industry teams to conduct long-term projects.
- *Interdisciplinary Grants in the Mathematical Sciences (IGMS), NSF 99-157,* enables mathematical scientists to undertake research and study in another discipline so as to expand their skills and knowledge to areas other than the mathematical sciences; to subsequently apply this knowledge to their research; and to enrich the educational experiences and broaden the career options of their students. Recipients are expected to spend 11 months full-time (within a 12-month period) either in a nonmathematical academic science department or in an industrial, commercial, or financial institution. These awards are in addition to those provided by GOALI.

In addition, program officers from different divisions/directorates can work together to coreview and potentially cofund cross-disciplinary proposals. For example, in 1992, an applied mathematician, a statistician, and an experimental biologist submitted a proposal called *Nonlinear Demographic Dynamics: Mathematical Models, Biological Experiments, and Data Analysis* to the Division of Mathematical Sciences (DMS). The principal program officer asked a co-worker from the Division of Environmental Biology to independently review the proposal. Neither review warranted traditional funding of the proposal. However, the DMS program officer thought that the cross-disciplinary research held promise and offered the researchers a very small grant to further develop the proposal. A year later, the group submitted a stronger proposal, which earned highly favorable reviews by both peer review groups. The updated proposal was funded at an amount 20 times that of the earlier developmental award. The research led to notable publications, including the article "Chaotic dynamics in an insect population," *Science* 275:389-391, which reports the observation of classical signs of chaos—period-doubling, bifurcations, and strange attractors—in laboratory populations of the flour beetle, providing rare and compelling data in support of a controversial theory of nonlinear population dynamics.[5]

The Office of Multidisciplinary Activities (OMA) is another resource available to program officers seeking supplementary funding for new, cross-disciplinary proposals. Located in the MPS directorate, this office will match or supplement divisional funds for a particularly novel, challenging, or complex multidisciplinary research project that does not fit well into the existing program structure or whose realization might otherwise be hampered by institutional and

---

[5]For additional information about this research, see the American Mathematical Society's publication *What's Happening in the Mathematical Sciences 1998-1999*, pp. 73-81.

procedural barriers. Although the OMA is located within the MPS, its funds are available to provide supplementary funding for crosscutting research involving a mathematical and/or physical science and other area of science. The OMA does not accept proposals directly, it does not review proposals, and it does not fund grant renewals.

Program Announcements

The *Overview of Programs*, at <www.nsf.gov/home/programs/start.htm>, lists programs by subject area. Specific announcements, identified according to the format NSF YY-NN(N), are filed at <www.nsf.gov/cgi-bin/getpub?nsfYYNN(N)>.

Review

The NSF Proposal Processing Unit assigned to the relevant NSF program inspects proposals before sending them to program officers for review. Each proposal then becomes the responsibility of a single program officer within a division of the NSF. Proposals are subject to a peer review process involving scientists and/or mathematical scientists who are not employed by the NSF. Peer review is done by mail, by convening a panel of experts, or both methods. This process is sometimes augmented by a site visit.

As of October 1998, reviewers have been asked to evaluate proposals according to two criteria: (1) the intellectual merit of the proposed activity and (2) the broader impacts of the proposed activity. For each criterion, the directions provided to reviewers suggest several questions to be considered:

Criterion 1: What is the intellectual merit of the proposed activity? How important is the proposed activity to advancing knowledge and understanding within its own field and across different fields? How well qualified is the proposer (individual or team) to conduct the project? (If appropriate, please comment on the quality of prior work.) To what extent does the proposed activity suggest and explore creative and original concepts? How well conceived and organized is the proposed activity? Is there sufficient access to resources?

Criterion 2. What are the broader impacts of the proposed activity? How well does the activity advance discovery and understanding while promoting teaching, training, and learning? How well does the proposed activity broaden the participation of underrepresented (from the standpoint of, say, gender, ethnicity, or geography) groups? To what extent will it enhance the infrastructure for research and education such as facilities, instrumentation, networks, and partnerships? Will the results be disseminated broadly to enhance scientific and technological understanding? What are the possible benefits of the proposed activity to society?

In addition, principal investigators should, where possible, indicate how the proposed research accomplishes two further goals of the NSF:

1. Integrating research and education by infusing education with the excitement of research and enriching research through the diversity of the learner perspectives and

2. Integrating diversity into NSF programs by broadening opportunities for minorities and other groups that are underrepresented in the sciences.

Review usually takes place at the division level: the division director makes funding decisions based on the recommendations of the program officers. As stated above, in the case of cross-disciplinary proposals, the principal program officer may ask a program officer from another division or directorate to coordinate an independent review of the proposal or to otherwise assist him in reviewing it.

Tips

The flexible nature of the funding process affords program officers significant discretion as to the manner in which cross-disciplinary proposals are evaluated (i.e., within a single division or through coreview). Thus, the means by which these proposals are evaluated vary across the NSF. Cross-disciplinary investigators should contact directors or program officers from the appropriate divisions prior to submitting proposals. If the feedback is positive, the researcher should, in the cover letter submitted with the proposal, specify that the research is cross-disciplinary and identify the other program officer(s) who have shown interest in the proposed research. The letter should request that the proposal be considered for cofunding across directorates and, where applicable, for OMA support. In addition, it should specify what contribution(s) the proposed research is expected to make to each discipline. (A copy of this letter should be sent to the program officer(s) identified therein.) If a proposal is not funded, the researcher(s) should seriously consider refining it and submitting it the following year.

Appeals

A grant applicant who believes his or her proposal was improperly reviewed can formally request that the associate director of the appropriate directorate investigate the means by which the proposal was evaluated. The associate director may perform the investigation or solicit someone from another division to handle the appeal.

Department of Energy
<www.doe.gov>

Science Programs

DOE programs of scientific research are organized under its Office of Science (formerly the Office of Energy Research). Program areas include the following:

- *Basic Energy Sciences,* which includes programs in materials science, chemical science, engineering, geosciences, and energy biosciences;
- *Fusion Energy Sciences,* which funds both domestic and international programs pursuing fusion energy through its International and Technology Division and its Science Division;
- *Biological and Environmental Research,* which includes life sciences research, medical sciences, and environmental sciences;
- *High Energy and Nuclear Physics;* and
- *Advanced Scientific Computing Research,* which includes mathematical, information, and computational sciences, and advanced energy projects and technology research.

Program Announcements

The Office of Science maintains a grants and contracts Web site at <http://www.er.doe.gov/production/grants/grants.html>, which posts solicitations inviting grant and contract applications. Recent areas of application have included plasma physics, environmental meteorology, human genome technology advances, environmental management science, energy biosciences, and Next Generation Internet. A guide to grants may be found at <http://www.er.doe.gov/production/grants/guide.html>.

Specific Programs

The Mathematical, Information, and Computational Sciences Division, <http:www.er.doe.gov/production/octr/mics/index.html>, has long supported applied mathematics relevant to the DOE's mission areas. For example, projects are ongoing in areas such as combustion modeling, materials processing, microscale modeling of materials, and photolithography and etching. Many of its grants link academic and industrial researchers with scientists or mathematical scientists at the national laboratories.

Numerous research and training opportunities exist at the DOE laboratories for university students at both the graduate and undergraduate level, for postdoctoral researchers, and faculty. Information on these programs can be found at <http://www.sandia.gov/ESTEEM/home.html>.

Information on the Computational Science Fellowship highlighted in Chapter 2 can be found at <http://www.krellinst.org/CSGF>.

Review

Project managers review applications for technical/scientific merit and program policy factors. In addition, the application is generally submitted to at least three qualified reviewers for evaluation. Such additional reviewers may be federal employees (including those from the Office of Science who are neither the selecting official nor in a direct line of supervision above the project manager) or nonfederal employees. All reviewers serve as advisors to the selecting official, and their recommendations are not binding. The Office of Science utilizes various types of review mechanism to accomplish a merit review: field readers, standing committees, and ad hoc committees. Upon request, applicants will be provided with a summary of the evaluation of their application. Selection of applications for award will be done by the authorized Office of Science selecting official and will be based upon merit review, the importance and relevance of the proposed project to Office of Science missions, and funding availability.

Tips

Office of Science policy encourages a potential applicant to discuss his or her proposed research project with Office of Science program staff to clarify areas of research interest before submitting the applications. In addition, an optional preapplication process allows the potential applicant to receive a response from the cognizant program office regarding the suitability of his or her proposed research project to DOE interests.

DEPARTMENT OF DEFENSE MISSION-ORIENTED AGENCIES

Defense Advanced Research Projects Agency
<www.darpa.mil>

Mission

The aim of the Defense Advanced Research Projects Agency (DARPA) is to develop imaginative, innovative, and often high-risk research ideas offering a significant technological impact that will go well beyond the normal evolutionary developmental approaches and to pursue these ideas from the demonstration of technical feasibility through the development of prototype systems.

Technical Offices and Their Missions

- The Advanced Technology Office (ATO) explores high-payoff programs in the areas of maritime, communications, and early entry/special forces operations. This is accomplished through development and transitioning of demonstrated systems for military users to respond to new and emerging threats.
- The Defense Science Office (DSO) mission is to identify and pursue the most promising technologies within the basic science and engineering research community and develop them into new DOD capabilities.
- The Information Systems Office (ISO) mission focuses on revolutionizing national security and military operations through the power of information systems technology...to know, to know more, to know faster, and be able to act flexibly.
- The Information Technology Office (ITO) focuses on inventing the networking, computing, and software technologies vital to ensuring DOD military superiority.
- The Microsystems Technology Office (MTO) mission focuses on the heterogeneous microchip-scale integration of electronics, photonics, and microelectromechanical systems (MEMS). Their high-risk, high-payoff technology is aimed at solving the national level problems of protection from biological, chemical, and information attack and providing operational dominance for mobile distributed command and control, combined manned/unmanned warfare, and dynamic, adaptive military planning and execution.
- The Special Projects Office (SPO) mission focuses on technologies that counter present and emerging national challenges in the areas of advanced detection and sensor systems, guidance and navigation capabilities, and underground facilities and unmanned aerial vehicles (UAV).
- The Tactical Technology Office (TTO) engages in high-risk, high-payoff advanced military research, emphasizing the "system" and "subsystem" approach to the development of aeronautic, space, and land systems, as well as embedded processors and control systems.

Applied and Computational Mathematics Program

DARPA's Applied and Computational Mathematics Program (ACMP) seeks to identify, develop, and demonstrate new mathematical paradigms enabling maximum performance at minimum cost in a wide variety of DOD systems applications. ACMP looks for opportunities to aggressively leverage the power of mathematical representations to effectively exploit the power of large-scale computational resources as they apply to specific problems of interest. ACMP's products are typically advanced algorithms and design methodologies.

ACMP is pursuing the development of well-conditioned fast algorithms and strategies for the exploitation of high-dimensional data (i.e., data with a high number of degrees of freedom) in order to deal with a variety of complex military problems such as adaptive array processing for missile seekers, waveform design for spaceborne sensors and communication applications, virtual integrated prototyping of advanced materials processing, efficient high-fidelity scattering computations for radar cross sections, and efficient mapping of signal-processing kernels onto advanced DOD hardware architectures.

There are currently four major R&D areas in DARPA's Defense Applications of Advanced Mathematics program: (1) signal and image processing, (2) virtual integrated prototyping of material processing, (3) fast and scalable scientific computation, and (4) virtual electromagnetic test range for air vehicles, <http://www.darpa.mil/dso/thrust/am/index.htm>.

Program Solicitations

Solicitations are published in the *Commerce Business Daily* and are listed, by office, at <www.darpa.mil/baa>.

Review

Owing to the proprietary nature of the solicited research, DARPA does not implement a peer review process.[6] The appropriate program manager (usually listed on the program solicitation) conducts a technical review of incoming proposals, and a final decision is made by the director of the appropriate technical office. To receive support, proposals must promise new and innovative mathematical science, science, or both, and the proposed research must further an application-oriented defense objective.

Programs can be proposal-driven. If a program manager receives a particularly novel proposal with the potential to open up a new area of research or sufficiently expand a current one, he will use discretionary funds to support it or will consider building a program around it. Although DARPA supports novel and bold research proposals, it does not usually provide start-up funding for projects in the very early stages of development.

---

[6]Instead of using advisory committees to review proposals, DARPA employs committees of experts to provide technical expertise and long-term program guidance. For example, DSO and ETO draw on the expertise of the Defense Sciences Research Committee (DSRC), which is run by an independent contractor.

Tips

Proposals should indicate what type of mission-related application the research is expected to have. Researchers are strongly encouraged to discuss their ideas with program managers before submitting proposals, particularly those that are unsolicited (i.e., not submitted in response to a broad agency amendment ((BAA)).

Office of Naval Research
<www.onr.navy.mil>

Science and Technology Departments

The Office of Naval Resesarch (ONR) solicits long-range science and technology (S&T) research projects in the following areas:

- Information, electronics, and surveillance: Electronics Division; Mathematical, Computer and Information Sciences Division; and Surveillance, Communications and Electronic Combat Division;
- Ocean, atmosphere, and space: Sensing and Systems Division and Processes and Prediction Division;
- Engineering, materials, and physical science: Physical Sciences S&T Division; Materials Sciences S&T Division; Mechanics and Energy Conservation S&T Division; Ship Structures and Systems S&T Division; Biomolecular and Biosystems S&T Division;
- Human systems: Medical S&T Division and Cognitive and Neural S&T Division;
- Naval expeditionary warfare operations technology: Strike Technology Division and Expeditionary Warfare Operations Technology Division; and
- Industrial and corporate programs: Manufacturing S&T (MANTECH) Division and Product Innovation Division.

ONR solicits the projects through the main broad agency announcement, 98-019, at <www.onr.navy.mil/02/baa/>.

Specific Programs

- *Faculty Sabbatical Leave Program,* at <www.onr.navy.mil/sci_tech/special/onrpgadr.htm>, allows faculty on sabbatical leave to conduct research at Navy laboratories. Appointments are for at least one semester but no more than a year.

- *U.S. Navy Summer Faculty Research Program,* at <www.onr.navy.mil/sci_tech/special/onrpgadn.htm>, gives sciences and engineering faculty members from institutions of higher learning the opportunity to participate in research at Navy laboratories for a 10-week period during the summer break. Three levels of appointment are available: summer faculty fellow, senior summer faculty fellow, and distinguished summer faculty fellow. Level of appointment is determined by a committee of scientists and engineers. Expert technical cooperation is needed in areas such as physics, chemistry, mathematics, electronics, aerodynamics, and materials science engineering (software, hardware, processing, etc.).

Review

ONR proposals are subject to scientific/technical review by program officers and further review by senior ONR managers. Proposals are not subject to peer review. Most proposals are evaluated according to the following criteria:

- Overall scientific, technical, and/or socioeconomic merits of the proposal;
- Potential naval relevance and contributions of the effort to the agency's specific mission;
- The principal investigator's capabilities, related experience, facilities, techniques, or unique combinations of these, which are integral factors for achieving the proposal objectives;
- The qualifications, capabilities, and experience of the proposed principal investigator, the team leader, and key personnel who are critical in achieving the proposal objectives; and
- Realism of the proposed cost and availability of funds.

For some programs, proposal submission is a two-step process: white papers are sought from prospective principal investigators. Based on the evaluation of these papers, selected applicants are invited to submit full proposals.

Tips

Investigators should identify which type(s) of application(s) potentially exist(s) for the proposed research and specify in the proposal/white paper how the proposed research ties into the broad mission of ONR. Before preparing a proposal/white paper, a potential principal investigator should contact the ONR program officer(s) whose program best matches his or her field(s) of interest. To identify these individuals, consult *Research Interests and Division Directors,* <www.onr.navy.mil/sci_tech/special/yip>.

Army Research Office
<www.aro.army.mil>

Areas of Interest

The Army Research Office (ARO) is interested in research in several areas: biosciences, chemistry, electronics, engineering sciences, environmental sciences, mathematical and computer sciences, materials sciences, and physics. These areas are described more fully in the broad agency announcement DAAD19-99-R-BAA1, which is intended to cover, in a general way, all research areas of interest to ARO. This document is published in the *Commerce Business Daily* and is available online at the ARO Web site under "Annual BAA."

Review

ARO reviews research proposals in two phases. Staff members perform initial reviews based on scientific merit and the potential contribution of the proposed research to the Army's mission. In addition, they determine if funds are available for the proposed effort. The surviving proposals are then subject to extensive peer review by scientists from within the government and by scientists and other experts outside the government.

Most proposals are evaluated using the criteria listed below (in descending order of importance):

- The overall scientific and/or technical merits of the proposal;
- The potential contributions of the effort to the ARO mission and the extent to which the research effort will contribute to balancing the overall ARO research program;
- The principal investigator's capabilities, related experience, facilities, techniques, or unique combinations of these which are integral factors for achieving the proposed objectives;
- The qualifications, capabilities, and experience of the principal investigator, team leader, or other key personnel who are critical to achievement of the proposed objectives;
- The proposed principal investigator's record of past performance; and
- The reasonableness and realism of the proposed costs and fee and the availability of funds.

Tips

Researchers contemplating submission of a proposal are encouraged to contact the appropriate ARO program manager/scientist to ascertain the extent of interest in the specific research project. Only a small number of comprehensive, interdisciplinary programs can be initiated in a single fiscal year. Inquiries about comprehensive and interdisciplinary programs should be sent to the director of ARO.

Prospective principal investigators are encouraged to submit proposals in two distinct phases:

- Phase I proposals provide the technical and budgetary information set forth in Section 3 of Part IV of the BAA.
- Phase II proposals (i.e., those that survive phase I) include the additional information identified in Section 4 of Part IV of the BAA.

Air Force Office of Scientific Research
<www.afosr.af.mil>

Research Interests

The Air Force Office of Scientific Research (AFOSR) is interested in research in the following areas:

- Aerospace and materials sciences

    Structural mechanics,
    Mechanics of composite materials,
    Unsteady aerodynamics and hypersonics,
    Turbulence and rotating flows,
    Combustion and diagnostics,
    Space power and propulsion,
    Metallic materials,
    Ceramics and nonmetallic materials, and
    Organic matrix composites.

- Physics and electronics

    Plasma physics,
    Space electronics,
    Atomic and molecular physics,
    Imaging physics,
    Optoelectronic information processing: devices and systems,
    Optical and photonic physics,
    Quantum electronic solids,
    Semiconductor materials, and
    Electromagnetic materials.

- Chemistry and life sciences

    Polymer chemistry,
    Surface science,
    Theoretical chemistry,
    Molecular dynamics,
    Chronobiology and neural adaptation,
    Perception and cognition,
    Sensory systems, and
    Bioenvironmental science.

- Mathematics and space sciences

> Dynamics and control,
> Physical mathematics and applied analysis,
> Computational mathematics,
> External aerodynamics and hypersonics,
> Optimization and discrete mathematics,
> Signals communication and surveillance,
> Software and systems,
> Artificial intelligence,
> Electromagnetics,
> Upper atmospheric physics, and
> Space sciences.

These areas are discussed in the broad agency accouncement, which can be acccessed from <http:ecs.ram.com/afosr/afr/afo/any/menu/any/afrfund.htm#research>.

Program Announcements

The Air Force solicits all research through broad agency announcements BAA-1 and specialized BAAs, which are published in the *Commerce Business Daily* and at the above Web site.

Review

Peer review, scientific/technical review by program officers, or both, are used to evaluate proposals. Unless otherwise stated, proposals are evaluated under the following two primary criteria, which are of equal importance:

1. The scientific and technical merits of the proposed research and
2. The potential contributions of the proposed research to the mission of the Air Force.

Other evaluation criteria used in the technical reviews, which are of lesser importance than the primary criteria and equal to one another in importance, are as follows:

3. The likelihood that the proposed effort will develop new research capabilities and broaden the research base in support of national defense;
4. The proposed principal investigator's, team leader's, or key personnel's qualifications, capabilities, related experience, facilities, or techniques or a combination of these factors that is integral to achieving Air Force objectives;
5. The past performance of the principal investigator and associated personnel; and
6. The realism and reasonableness of proposed costs and availability of funds; although not a primary evaluation factor, price is a substantial factor in the selection of proposals for award.

Tips

Proposals should clearly indicate how the proposed research ties into the mission and research interests of AFOSR as described in *AFOSR Research Interests Brochure and Broad Agency Announcement*.

Prospective principal investigators are encouraged to consult with AFOSR technical personnel for opinions on any proposed research before completing and submitting a full proposal. Sometimes a preliminary proposal is submitted. It should be in letter format and briefly describe the proposed research project's (1) objective, (2) general approach, and (3) impact on DOD and civilian technology, as well as any unique capabilities or experience of the principal investigator (e.g., collaborative research activities involving Air Force, DOD, or other federal laboratories). Preproposal letters should not exceed three typewritten pages.

E

Acronyms and Abbreviations

ACMP	applied computational mathematics programs (DOD)
AHCs	academic health centers
AMS	American Mathematical Society
ASM	American Society for Microbiology
BAA	broad agency announcement
BMS	Board on Mathematical Sciences (NRC)
CAM	Center for Scientific Applications of Mathematics
CDC	Centers for Disease Control and Prevention
CNRS	Centre National de la Recherche Scientifique
CSGF	Computational Science Graduate Fellowship
CSR	Center for Scientific Review (NIH)
CT	computer-assisted tomography
DARPA	Defense Advanced Research Projects Agency
DIMACS	Center for Discrete Mathematics and Theoretical Computer Science
DOD	Department of Defense
DOE	Department of Energy
DSO	Defense Sciences Office (DOD)
ETO	Electronics Technology Office (DOD)
GFD	Geophysical Fluid Dynamics Program (WHOI)
HPCC	high-performance computing and communications
IGERT	Integrative Graduate Education and Research Training
IMA	Institute for Mathematics and Its Applications
IOM	Institute of Medicine
IRG	initial review group
JPBM	Joint Policy Board for Mathematics
MERC	Molecular Engineering and Research Center
MRI	magnetic resonance imaging
MSED	Mathematical Sciences Education Board (NRC)
MSRI	Mathematical Sciences Research Institute
NAE	National Academy of Engineering
NAS	National Academy of Sciences
NCSA	National Center for Supercomputing Applications
NIGMS	National Institute of General Medical Sciences (NIH)
NIH	National Institutes of Health
NIST	National Institute of Standards and Technology
NRC	National Research Council
NSF	National Science Foundation
OMA	Office of Multidisciplinary Activities (NSF)
ONR	Office of Naval Research
PET	positron emission tomography

PMMB	Program in Mathematics and Molecular Biology
RFA	request for application
SIAM	Society for Industrial and Applied Mathematics
SPE	special emphasis panel (NIH)
SPECT	single photon emission tomography
SPRT	Sequential Probability Ratio Test
SRA	scientific review administrator
STCs	science and technology centers (NSF)
TICAM	Texas Institute for Computational and Applied Mathematics
WHOI	Woods Hole Oceanographic Institution